OKANAGAN UNIV/COLLEGE LIBRARY

P9-BHY-711

The Second Law

THE SECOND LAW

P. W. Atkins

SCIENTIFIC AMERICAN LIBRARY

Scientific American Books
An imprint of W. H. Freeman and Company
New York

Library of Congress Cataloguing in Publication Data:

Atkins, P. W. (Peter William), 1940–
　The second law.

　Bibliography: p.
　Includes index.
　1. Thermodynamics.　I. Title
QC311.5.A8　1984　536'.71　84-5377
ISBN 0-7167-6006-1

Copyright © 1984, 1994 by P. W. Atkins

No part of this book may be reproduced by any
mechanical, photographic, or electronic process,
or in the form of a phonographic recording,
nor may it be stored in a retrieval system,
transmitted, or otherwise copied for public or
private use, without written permission from the
publisher.

Printed in the United States of America

Book design by Malcolm Grear Designers

Scientific American Library is published by
Scientific American Books, Inc., a subsidiary
of Scientific American, Inc.

Distributed by W. H. Freeman and Company,
41 Madison Ave., New York, New York 10010

1 2 3 4 5 6 7 8 9 0　KP　9 9 8 7 6 5 4

CONTENTS

1   **NATURE'S DISSYMMETRY**   **1**
The identification of energy   3
The laws of thermodynamics   8
Revolutions of dissymmetry   10
The identification of the dissymmetry   14

2   **THE SIGNPOST OF CHANGE**   **23**
The nature of heat and work   23
The seeds of change   24
Towards corruption   29
Entropy   30
Measuring the entropy   35
The dissipation of quality   37
Ceilings to efficiency   39
The end of the external   42

3   **COLLAPSE INTO CHAOS**   **45**
Inside energy   46
Modeling the Universe   50
Temperature   55
The direction of natural change   57
Natural processes   62

4   **THE ENUMERATION OF CHAOS**   **65**
Boltzmann's Demon   66
The Demon's cage   72
Chaos, coherence, and corruption   75

5   **THE POTENCY OF CHAOS**   **81**
Carnot under the microscope   82
Stirling's engine   87
Internal combustion   94
Turbine power   102
Towards coherence   105

6   **TRANSFORMATIONS OF CHAOS**   **107**
Chemical transformations   108
Iron burning   111
Cooling as heating   114
The rate of dispersal   121
Chaos and order   124

7 POWERS OF TEMPERATURE 127
Normal life   129
Catching cold   130
The first power down   137
The second power down   139
Lower powers down   142
Powers hotter   147
Hotter than infinity   149
Toward life   155

8 CONSTRUCTIVE CHAOS   157
The emergence of intricate form   157
Proteins   160
Free energy   165
The unnatural reactions of life   172
The electrochemistry of life   175

9 PATTERNS OF CHAOS   179
Structure   179
Dissipative structures   183
The emergence of complexity   189
The apotheosis of the steam engine   198

APPENDIX 1: UNITS   201

APPENDIX 2: FORMALITIES   203
Thermodynamics   203
Temperature   207

SOME SOURCES FOR FURTHER READING   208

SOURCES OF ILLUSTRATIONS   210

INDEX   211

PREFACE

No other part of science has contributed as much to the liberation of the human spirit as the Second Law of thermodynamics. Yet, at the same time, few other parts of science are held to be so recondite. Mention of the Second Law raises visions of lumbering steam engines, intricate mathematics, and infinitely incomprehensible entropy. Not many would pass C. P. Snow's test of general literacy, in which not knowing the Second Law is equivalent to not having read a work of Shakespeare.

In this book I hope to go some way toward revealing the workings of the Law, and showing its span of application. I start with the steam engine, and the acute observations of the early scientists, and I end with a consideration of the processes of life. By looking under the classical formulation of the Law we see its mechanism. As soon as we do so, we realize how simple it is to comprehend, and how wide is its application. Indeed, the interpretation of the Second Law in terms of the behavior of molecules is not only straightforward (and in my opinion much easier to understand than the First Law, that of the conservation of energy), but also much more powerful. We shall see that the insight it provides lets us go well beyond the domain of classical thermodynamics, to understand all the processes that underlie the richness of the world.

One feature of this account is to establish that the steam engine, from which our knowledge of the Second Law emerged, is a device that models in the clearest way the basic irreversibility of the world. The nineteenth-century physicists who cut away the technological fog around the engine, and perceived the principles of change in all its forms, were able to do so because the steam engine embodies the central feature of all change in a very simple way. Then, with that insight established, it becomes possible to find ever more subtle examples in which the underlying simplicity is shrouded, and then to unravel the shroud until we see the simplicity again. Thus we can trace back biological events to processes exhibited in stark simplicity by the steam engine. We shall make that journey from the steam engine to the brink of consciousness.

One difficulty with accounts of thermodynamics, and particularly of the Second Law in either its classical or its statistical form, is that the subject is intrinsically mathematical. In the pages that follow, I try to avoid all mathematics, and though there is the occasional equation when a particular point has to be made, I assume very little scientific background. Except for a few places, this account should therefore be accessible to any persistent reader, however scientifically unprepared. To the scientifically informed I offer my apologies for a slowness of pace here and there, but I hope that even such a reader will be rewarded at other stages in the text with a new angle on what is going on in the world.

I am conscious of a major omission in the material I present: I have deliberately omitted reference to the relation between *information theory* and entropy. On the one hand, I agree that the principles and mathematics of information theory can contribute substantially to the formulation of thermodynamics and the expression of its content. On the other hand, there is the danger, it seems to me, of giving the impression that entropy requires the existence of some cognizant entity capable of possessing "information" or of being to some degree "ignorant." It is then only a small step to the presumption that entropy is all in the mind, and hence is an aspect of an observer. I have no time for this kind of muddleheadedness, and intend to keep such metaphysical accretions firmly at bay. For this reason I omit any discussion of the analogies between information theory and thermodynamics.

The structure of the discussion is as follows. From observations on the steam engine (Chapter 1), we watch the early thermodynamicists distill the Second Law (Chapter 2). Then we plunge inside matter (Chapter 3), and see that at the level of the behavior of particles the Second Law has a simple and richly fruitful formulation. Next (in Chapter 4) we explore the extent to which this qualitative insight can be rendered quantitative, and it is here that a little mathematics obtrudes (but it is not crucial to the remainder of the book). With the underlying ideas established, we return (in Chapter 5) to the steam engine and its descendants, and see how they account for the conversion of heat into work. With the production of work explained, we turn to the production of matter (Chapter 6), and see that the ideas that lie beneath physics account for chemistry too. One idea that is gradually elaborated throughout the text is the idea of *structure*. When we have established the role that the Second Law plays in simple physics and chemistry, we see how structural order can be imposed in a variety of ways: by physical change, especially by refrigeration (Chapter 7) and by chemical change (Chapter 8). In Chapter 8 we also see how the Second Law accounts for the emergence of the intricately ordered forms characteristic of life. Then finally (Chapter 9) we stand back and look at how far we have traveled, come to grips with a deep version of the idea of "structure," and see how it emerges from the rule of the Second Law. There are a few Appendixes: one to explain units of energy and some related information;

a second to give a whirlwind tour of some of the formal thermodynamics lying in the background of the discussion.

I have long had an interest in the Second Law, but two meetings helped to bring my thoughts into focus. One was organized by Aleksandra Kornhauser in Yugoslavia, the second by George Marx in Hungary. Max Whitby of the B.B.C. sharpened my appreciation of some of the games we play. Alexandra MacDermott and John Rowlinson of the University of Oxford and Philip Morrison of M.I.T. spent much time in providing judicious and helpful comments. Aidan Kelly worked hard to ensure that I knew what he thought I wanted to say. To all these I am grateful.

P. W. Atkins
Oxford, England
January 1984

The Second Law is timeless, and I would like to think that *The Second Law* is timeless, too. But, given the opportunity, I did make a few adjustments for this republication in paperback. I have brought various items up to date in the text, changed some of the illustrations and their legends to improve clarity, and replaced a few photographs to improve the appearance of the text. I have also substantially revised the Further Reading section, mentioning a number of current texts as well as the classics. The BASIC programs in the Appendices that seemed almost before their time in 1984 have become museum pieces, and have been deleted. The theme of the book—the extraordinary creative potency of decay into disorder—remains intact, as does the motivation of the presentation—to explain an intrinsically mathematical subject of great importance in prose and pictures.

P. W. Atkins
Oxford, England
May 1994

The Second Law

1 NATURE'S DISSYMMETRY

Nicolas Leonard Sadi Carnot (1796–1832).

War and the steam engine joined forces and forged what was to become one of the most delicate of concepts. Sadi Carnot, the son of a minister of war under Napoleon and the uncle of a later president of the Republic, fought in the outskirts of Paris in 1814. In the turmoil that followed, he formed the opinion that one cause of France's defeat had been her industrial inferiority. The contrast between England's and France's use of steam epitomized the difference. He saw that taking away England's steam engine would remove the heart of her military power. Gone would be her coal, for the mines would no longer be pumped. Gone would be her iron, for, with wood in short supply, coal was essential to ironmaking. Gone, then, would be her armaments.

But Carnot also perceived that whoever possessed efficient steam power would be not only the industrial and military master of the world, but also the leader of a social revolution far more universal than the one France had so recently undergone. Carnot saw steam power as a universal motor. This motor would displace animals because of its greater economy, and would supersede wind and water because of its reliability and its controllability. Carnot saw that the universal motor would enlarge humanity's social and economic horizons, and carry it into a new world of achievement. Many people today can see the early steam engines, those cumbersome hulks of wood and iron, only as ponderous symbols of the squalor and poverty that typified the newly emerging industrial society. In fact, those earthbound leviathans proved to be the wings of humanity's aspirations.

Carnot was a visionary and a sharp analyst of what was needed to improve the steam engine (as his father had been an acute analyst of mechanical devices), but he could have had no inkling of the *intellectual* revolution to which his technologically motivated studies would lead. In discovering that there is an intrinsic inefficiency in the conversion of heat into work, he set in motion an intellectual mechanism which a century and a half later embraces all activity. In pinning down the efficiency of the steam engine and circumscribing its limitations, Carnot was unconsciously establishing a new attitude toward all kinds of change, toward the conversion of

One of the earliest steam engines. Its analysis stimulated the ideas we explore in this book.

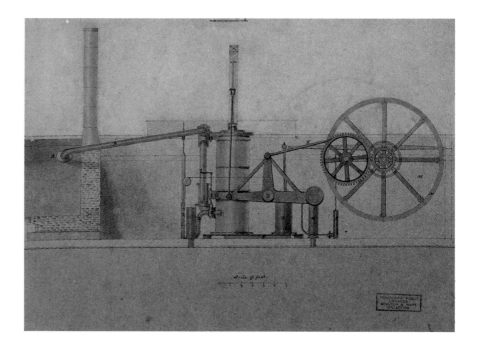

the energy stored in coal into mechanical effort, and even toward the unfolding of a leaf. Moreover, he was also establishing a science that went beyond the apparently abstract physics of Newton, one that could deal with both the abstractions of single particles and the reality of engines. All this encapsulates the span of topics in this book: we shall travel from the apparently coarse world of the early industrial engine to the delicate and refined world of the enjoyment of beauty, and in doing so we shall discover them one.

Carnot's work (which was summarized in his *Refléxions sur la puissance motrice du feu*, published in 1824) was based on a misconception; yet, even so, it laid the foundations of our subject. Carnot subscribed to the then-conventional theory that heat was some kind of massless fluid or *caloric*. He took the view that the operation of a steam engine was akin to the operation of a water mill, that caloric ran from the boiler to the condenser, and drove the shafts of industry as it ran, exactly as water runs and drives a mill. Just as the quantity of water remains unchanged as it flows through the mill in the course of doing its work, so (Carnot believed) the quantity of caloric remained unchanged as it did its work. That is, Carnot based his analysis on the assumption that the quantity of heat was *conserved*, and that work was generated by the engine because the fluid flowed from a hot, thermally "high" source to a cold, thermally "low" sink.

The intellectual effort needed to disentangle the truth from this misconception had to await a new generation of minds. Among the generation born around 1820, there were three people who would take up the challenge and resolve the confusion.

The Identification of Energy

The first of these three was J. P. Joule, born in 1818. Joule was the son of a Manchester brewer. His wealth, and the brewery's workshops, gave him the opportunity to follow his inclinations. One such inclination was to discover a general, unifying theme that would explain all the phenomena then exciting scientific attention, such as electricity, electrochemistry, and the processes involving heat and mechanics. His careful experiments, done in the 1840s, confirmed that heat was *not* conserved. Joule showed by increasingly precise measurements that work could be converted *quantitatively* into heat. This was the birth of the concept of the *mechanical equivalence of heat*, that work and heat are mutually interconvertible, and that heat is not a substance like water.

James Prescott Joule (1818–1889).

William Thomson, Lord Kelvin (1824–1907).

Such was the experimental evidence that upset the basis for the conclusions, but not the conclusions themselves, that Carnot had drawn a generation before. Now it was time for the theoreticians to take up the challenge and to resolve the nature of heat.

William Thomson was born in Belfast in 1824, moved to Glasgow in 1832, and entered the university there at age ten, already displaying the intellectual vigor that was to be the hallmark of his life. Although primarily a theoretician, he had great practical ability. Indeed, his wealth sprang from a practical talent, which he polished by a brief sojourn in Paris after he graduated from Cambridge, where he had gone in 1843. His career at Glasgow resumed in 1846, when at 22 he was appointed to the chair of natural philosophy. He divided his time between theoretical analysis of the highest quality and moneymaking of enviable proportions from his work in telegraphy. Great Britain's preeminence in the field of international communication and submarine telegraphy can be traced to Thomson's analysis of the problems of transmitting signals over great distances, and his invention (and patenting) of a receiver that became the standard in all telegraph offices.

William Thomson, as is sometimes the confusing habit of the British, later matured into Lord Kelvin, by which name we shall refer to him from now on. His wealth and his practical attainments have now been largely forgotten. What remains as his lasting memorial, apart from a slab in Westminster Abbey, is his intellectual achievement.

Kelvin and Joule met at the Oxford meeting of the British Association for the Advancement of Science in 1847. From that meeting Kelvin returned with an unsettled mind. He was reported as being astounded by Joule's refutation of the conservation of heat. Although impressed with what Joule had been able to demonstrate, he believed that Carnot's work would be overthrown if heat were not conserved, and if there were no such thing as caloric fluid.

Kelvin began by setting forth the conceptual tangle that appeared to be confronting physics. He went on to develop the view (published in his paper "On the dynamical theory of heat" in 1851) that perhaps *two* laws were lurking beneath the surface of experience, and that in some sense the work of Carnot could survive without contradicting the work of Joule. Thus emerged the study, and the name, of *thermodynamics*, the theory of the mechanical action of heat, and the beginnings of the realization that Nature had two pivots of action.

The third significant mind born in the decade of the 1820s was that of Rudolf Clausius. Clausius was born in 1822. There should be nothing surprising in the fact that these three shapers of thermodynamics were contemporaries. Thermodynamics was the object of intellectual ferment of the time, and bright minds are attracted to bright possibilities. Clausius's first contribution cut closer to the bone than had Kelvin's. In dealing with the theme inspired by Carnot, carried on by Joule, and extended by Kelvin, in a monograph that was titled *Über die bewegende Kraft der Wärme* when it was published in 1850, Clausius sharply circumscribed the problems then facing thermodynamics, and in doing so made them more open to analysis. His was the focusing mind, the microscope to Kelvin's cosmic telescope.

Clausius also saw that the case of Carnot *vs.* Joule could to some extent be resolved if there were two underlying principles of Nature. He refined Carnot's principle, and rid the world of caloric, but he went further: although he carefully insulated his general conclusions from his speculations, he did go on to speculate on how heat could be explained in terms of the behavior of the particles of which matter is composed. That was the dawn of the modern era of thermodynamics.

Carnot was born in 1796 and died of cholera in 1832; by then he had let slip his belief in the reality of caloric. Joule, Kelvin, and Clausius were born in the period 1818–1824, and their generation thrust thermodynamics onto the intellectual stage. But it needed a third generation to unify this new discipline, and to attach it to the other currents of science which by then were starting to flow.

Ludwig Boltzmann was born in 1844. His contribution was to forge the link between the properties of matter in bulk, then being established by the deployment of Kelvin's and Clausius's thermodynamics, and the behavior of matter's individual particles, its atoms. Kelvin, Clausius, and their contemporaries developed the seed planted by Carnot, and were able to establish a great warehouse of relations between observations. However, comprehension of these relations could come only when a *mechanistic* explanation in terms of particles and their properties had been established.

Boltzmann perceived that identifying the cooperation between atoms which showed itself to the observer as the properties of bulk matter would take him into the innermost workings of Nature. Though short-sighted, he saw further into the workings of the world than most of his contemporaries, and he began to discover the deep structure of change; furthermore, he did all this before the existence of atoms was generally accepted. Many of his contemporaries doubted the credibility of his assumptions and his argument, and feared that his work would dethrone the purposiveness which they presumed to exist within the workings of the deeper world of change, just as Darwin had recently dispossessed its outer manifestations. Suffering from their scorn, Boltzmann was overcome by instability and unhappiness and killed himself.

Rudolf Clausius (1822–1888).

Ludwig Boltzmann (1844–1906), at age 60.

In 1906, when Boltzmann died, ideas were in the air, and techniques were becoming available, that were to win over his critics and to establish his reputation as one of the greatest of theoretical physicists. The emergence of quantum theory, together with the experimental exploration and detailed mapping of the structures of atoms, brought to the microscopic world a reality that, although out of joint with the familiar, was compelling and essentially indisputable. When that had been achieved, no one could seriously deny the existence of atoms, even though they appear to behave in a manner that at first sight (and still to some) seemed strange. Now we have techniques that show both individual atoms (below) and atoms strung together into molecules (on right). The fundamental basis of Boltzmann's viewpoint has been established beyond reasonable doubt, even though that microscopic world is far more peculiar than even he envisaged.

A scanning tunneling microscope image of gallium and arsenic atoms in the semiconductor gallium arsenide. The gallium atoms are green and the arsenic atoms are orange.

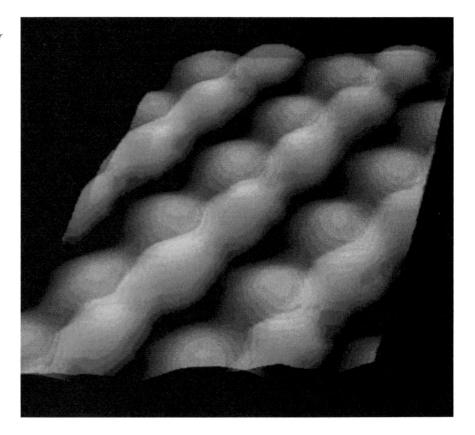

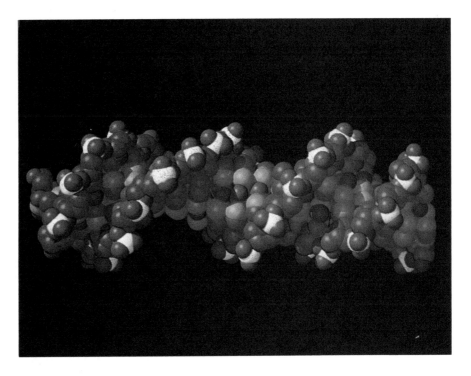

A computer-generated image of a fragment of DNA, the genetic coding molecule in the nuclei of cells.

The aims adopted and the attitudes struck by Carnot and by Boltzmann epitomize thermodynamics. Carnot traveled toward thermodynamics from the direction of the engine, then the symbol of industrialized society: his aim was to improve its efficiency. Boltzmann traveled to thermodynamics from the atom, the symbol of emerging scientific fundamentalism: his aim was to increase our comprehension of the world at the deepest levels then conceived. Thermodynamics still has both aspects, and reflects complementary aims, attitudes, and applications. It grew out of coarse machinery; yet it has been refined to an instrument of great delicacy. It spans the whole range of human enterprise, covering the organization and deployment of both resources and ideas, particularly ideas about the nature of change in the world around us. Few contributions to human understanding are richer than this child of the steam engine and the atom.

The Laws of Thermodynamics

The name *thermodynamics* is a blunderbuss term originally denoting the study of heat, but now extended to include the study of the transformations of energy in all its forms. It is based on a few statements that constitute succinct summaries of people's experiences with the way that energy behaves in the course of its transformations. These summaries are the *Laws of thermodynamics.* Although we shall be primarily concerned with just one of these laws, it will be useful to have at least a passing familiarity with them all.

There are *four* Laws. The third of them, the *Second Law,* was recognized first; the first, the *Zeroth Law,* was formulated last; the *First Law* was second; the *Third Law* might not even be a law in the same sense as the others. Happily, the content of the laws is simpler than their chronology, which represents the difficulty of establishing properties of intangibles.

The *Zeroth Law* was a kind of logical afterthought. Formulated by about 1931, it deals with the possibility of defining the temperature of things. Temperature is one of the deepest concepts of thermodynamics, and I hope this book will sharpen your insight into its elusive nature. As time is the central variable in the field of physics called dynamics, so temperature is the central variable in thermodynamics. Indeed, there are several amusing analogies between time and temperature that go deeper than the accidents that they both begin with and are represented by the same letter. For now, however, we shall regard temperature as a refinement and quantitative expression of the everyday notion of "hotness."

The *First Law* is popularly stated as "Energy is conserved." That it is *energy* which is conserved, not heat, was the key realization of the 1850s, and the one that Kelvin and Clausius presented to the world. Indeed, the emergence of energy as a unifying concept was a major achievement of nineteenth-century science: here was a truly abstract concept coming into a dominant place in physics. Energy displaced from centrality the apparently more tangible concept of "force," which had been regarded as the unifying concept ever since Newton had shown how to handle it mathematically a century and a half previously.

Energy is a word so familiar to us today that we can hardly grasp either the intellectual Everest it represents or the conceptual difficulty we face in saying exactly what it means. (We face the same difficulty with "charge" and "spin" and other fundamental familiarities of everyday language.) For now, we shall assume the concept of energy is intuitively obvious, and is conveyed adequately by its definition as "the capacity to do work." The shift in the primacy of energy can be dated fairly accurately. In 1846 Kelvin was arguing that physics was the science of force. In 1847 he met and listened to Joule. In 1851 he adopted the view that, after all, physics was the science of energy. Although forces could come and go, energy was here to stay. This concept appealed deeply to Kelvin's religious inclinations: God, he could now argue, endowed the world at the creation with a store of energy, and that divine gift would persist for eternity, while the ephemeral forces danced to the music of time and spun the transitory phenomena of the world.*

Kelvin hoped to raise the concept of energy beyond what it was becoming in the hands of the mid-nineteeth-century physicists, a mere constraint on the changes that a collection of particles could undergo without injection of more energy from outside. He hoped to establish a physics based solely on energy, one free of allusions to underlying models. He had a vision that all phenomena could be explained in terms of the transformations of energy, and that atoms and other notions were to be regarded merely as manifestations of energy. To some extent modern physics appears to be confirming his views, but it is doing so in its typical slippery way: without doing away with atoms!

The *Second Law* recognizes that there is a fundamental dissymmetry in Nature: the rest of this book is focused on that dissymmetry, and so we shall say little of it here. All around us, though, are aspects of the dissymmetry: hot objects cool, but cool objects do not spontaneously become hot; a bouncing ball comes to rest, but a stationary ball does not spontaneously begin to bounce. Here is the feature of Nature that both Kelvin and Clausius disentangled from the conservation of energy: although the total *quantity* of energy must be *conserved* in any process (which is their revised version of what Carnot had taken to be the conservation of the quantity of caloric), the *distribution* of that energy changes in an *irreversible* manner. The Second Law is concerned with the natural direction of change of the distribution of energy, something that is quite independent of its total quantity.

The *Third Law* of thermodynamics deals with the properties of matter at very low temperatures. It states that we cannot bring matter to a temperature of absolute zero in a finite number of steps. As I said earlier, the Third Law might not be a true Law of thermodynamics, because it seems to assume that matter is atomic, whereas the other Laws are summaries of

---

* A mischievous cosmologist might now turn this argument on its head. One version of the Big Bang, the *inflationary scenario,* can be interpreted as meaning that the total energy of the Universe is indeed constant, but constant at zero! The positive energy of the Universe (largely represented by the energy equivalent of the mass of the particles present, that is, by the relation $E = mc^2$) might exactly balance the negative energy (the gravitational attractive potential energy), so that overall the total might be zero. Thus Kelvin's God may have left a nugatory legacy.

direct experience and are independent of any such assumption. There is thus a difference of kind between this Law and the others, and even its logical implications seem less securely founded than theirs. We shall touch on it again, but only much later.

These, then, in broad and indistinct outline, are the Laws that stake out our domain: we have identified the territory; we shall proceed to explore its details. The Laws present us, however, with an immediate problem: thermodynamics is an intrinsically mathematical subject. Clausius's remarkably elegant *functional thermodynamics* is a collection of mathematical relations between observations; but with the relations gone, so too is the subject. Boltzmann's beautiful *statistical thermodynamics* (some of which is carved on his tombstone) also consists of its equations; without them, we have no subject to explore. That the subject is inherently mathematical is the principal reason why it remains so daunting.

Nevertheless, the subject is so important, and the implications of the Second Law so profound and far-reaching, that it seems worth the effort to discover a loophole in its mathematical defenses. What we shall do in the following pages, therefore, is attempt to explore thermodynamics without the mathematics. Then we shall not have the pain (a pain that many rightly regard as at least half the pleasure) of the mathematics. Although we shall necessarily remain outside the subject itself, we shall be able to share the insights it provides into the workings of the world.

But shall we remain so very much outside? Shall we be merely outsiders, tourists, while the real activities go on inside? A more optimistic attitude (and one applicable to other fields as well) is to take the view that mathematics is only a guide to understanding, a refiner of arguments and a purifier of comprehension, and not the endpoint of explanation. If that is so, then the people within are the unlucky toilers who are merely working to sharpen our wits. Whichever position you adopt, I hope the following pages will add something to your view of the world.

Revolutions of Dissymmetry

An intrinsic dissymmetry of Nature is reflected in our technological history. The conversion of stored energy and of work into heat has been commonplace for thousands of years. However, the widespread mastery of the reverse, the controlled conversion of heat and stored energy into work, dates principally from the industrial revolution. I say "principally" because work has, of course, been achieved for centuries. The conversion of wind—which is essentially a store of energy supplied by the Sun—into the motion of mills and ships is one example of such a conversion. The use of animals is another, even more indirect procedure with the same overall result. But we can regard the industrial revolution as the surge of activity released by humanity's sudden discovery of how to exploit energy, how to convert heat into work at will, so that changes in society were no longer limited by using animals to do work and by the one-sided processes of Nature.

Primitive people learned to produce *heat* at will and in abundance by burning fuels. Then, apart from reliance on such natural sources as winds and oxen, it took people thousands of years to discover the much more sophisticated procedures by which the energy in fuels could be converted into work (other than by feeding the fuels as food to cattle, horses, and slaves). The founders of the industrial revolution mastered the production of *work* in abundance and at will.

The differences in the degrees of sophistication needed to produce heat, on the one hand, and work, on the other, from the same fuel are apparent as soon as we look at the equipment each process requires. In order to produce heat from a fuel, all we need is an open hearth (below), on which the unconstrained combustion of the fuel—wood, coal, or animal

The combustion of matter, either accidental or intentional, is a primitive way of releasing the energy stored in fuels. This prairie brushfire in Bolivia releases energy as it burns.

On the other hand, a jet engine, which extracts the energy of fuels as work, is much more complicated.

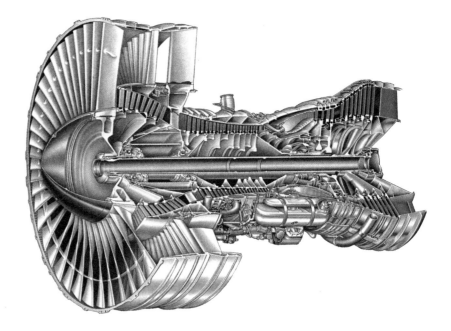

and vegetable oils—produces more or less abundant heat. In order to achieve *work,* we need a much more complicated device (above). Primitive people used the heat of their simple hearths to unlock the elements from the Earth, and from them gradually built the basis of civilization.

Although early minds were unaware of it, what their fires were releasing was the energy trapped from the Sun. (It is fitting, but coincidental, that many should have worshipped the Sun too.) At first the demands of civilization were slight, and could be satisfied by the energy which the Sun had shone down in recent years, and which had been stored in the annual growth of vegetation. But as civilization progressed, trapped solar energy that had been accumulated in former ages was increasingly exploited, and wood gave way to coal as the principal fuel. Nevertheless, this was not a technological revolution, for all that was happening was that people were mining further back in time, and retrieving the energy trapped from the Sun in an earlier epoch.

Modern civilizations continue this quest to mine the past for the harvest laid down in earlier times. Now we exploit the great stores of oil, the partially decayed remnants of marine life (which also drew its initial sustenance from the Sun). But such are our demands that we have been forced to dig beyond that time and collect the harvest of other stars. For example, the atoms of uranium we now burn in the complex hearths of nuclear reactors are the rich ashes of former stars. These atoms were formed in the death throes of early generations of stars, when light atoms were hurled against each other with such energy that they fused into progressively heavier ones. The old stars exploded, sloughing off the atoms and spreading them through space, to go through other roastings, explosions, and dispersals, until in due course they collected in the ball of rock we stand on and mine today.

But the quest for fire from the past goes on even deeper. Now we seek to mine beyond the formation of the Earth, beyond the deaths of generations of stars, and into the ash of the creation itself.

In the earliest moments of the Universe, the Big Bang shook spacetime to its foundations, and conditions of almost inconceivable tumult raged through the swelling cosmos; yet this great cataclysm managed to produce only the simplest atoms of all. The labor of the cosmic elephant resulted in the birth of a cosmic mouse: out from the tumult dropped hydrogen with a dash of helium. These elements, still superabundant, are the ashes of the Big Bang, and our attempts to achieve the controlled fusion of hydrogen into helium are aimed at capturing the energy they still store. Hydrogen is the oldest fossil fuel of all: when we master fusion, we shall be mining at the beginning of time.

The emergence and flourishing of civilizations has thus been characterized by our mining progressively further into the past for convenient, concentrated supplies of energy. Mining deeper in time, however, is merely an elaboration of the primitive discovery that energy can be unleashed as heat. However sophisticated the hearth, the combustion of fossils, whether of vegetation, stars, or the Big Bang, is merely a linear series of refinements of the basic discovery of combustion. Such refinements are not in themselves revolutions: they are sophistications—qualitative extensions—of processes that are almost as old as the hills.

Without the revolution that comes about from exploring the other side of Nature's dissymmetry, the conversion of heat into work, we would merely be warmer, not wiser. This other side lets us tap the store of energy in fuel and extract from it *motive power*. Then, with motive power we can make artifacts, we can travel, and we can even communicate without traveling. Why, though, did this dissymmetry take so long to exploit?

The task confronting humanity was to find a way to extract *ordered motion* from *disordered motion*, for therein lies the difference between heat and work. This is the moment when we must look more closely at the nature of the dissymmetry, and bring ourselves forward from the time before Carnot to the comprehension that came with Clausius and Kelvin.

The Identification of the Dissymmetry

We shall use a steam engine to identify Nature's dissymmetry. This is essentially what Carnot did. We shall then step inside the engine, so to speak, and discover the atomic basis of the dissymmetry of events. That is what Clausius identified and Boltzmann developed.

An engine is something that converts heat into work. Work is a process such as raising a weight (below). Indeed, we shall define *work* as *any process that is equivalent to the raising of a weight*. Later, as this story develops, we shall use our increased insight to build more general definitions and find the most all-embracing definition right at the end. That is one of the delights of science: the more deeply a concept is understood, the more widely it casts its net. Heat we shall come to later.

Work *is a way of transferring energy between a system and its surroundings; it is a transfer effected in such a way that a weight could be raised in the surroundings as a result. When work is done* on *a system, the change in the surroundings is equivalent to the lowering of a weight.*

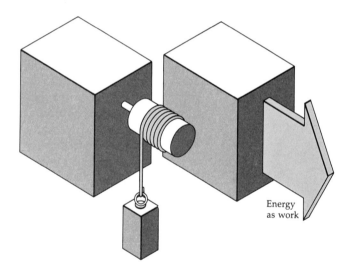

Energy
as work

An engine should be capable of operating indefinitely, and to go on making the conversion for as long as the factory operates or for as long as the journey lasts. Single-shot conversions, such as the propulsion of a cannonball by the combustion of a charge of powder, produce work, but are not engines in this sense. An engine is a device that operates *cyclically*, and returns periodically (once in each revolution, or once in several revolutions, of a crankshaft, for instance) to its initial condition. Then it can go on, in principle, for ever, living off the energy supplied by the hot source, which in turn is supplied with energy by the burning fuel.

Engines and the cycles they go through in the course of their operation may be as intricate as we please. A sequence of steps known as the *Carnot cycle* is a convenient starting point. The cycle is an abstract idealization, and very simple. Nevertheless, it may be elaborated (as we shall see later) to reproduce the stages that real engines such as gas turbines and jet engines go through, and it captures the essential feature of all engines. The engine itself (as illustrated below) consists of a gas trapped in a cylinder fitted with a piston. The cylinder can be put in contact with a hot source (steam from a boiler) and with a cold sink (cooling water), or it may be left completely insulated. Note that the operation does not capture the actual working of a steam engine, because in the Carnot engine steam is not admitted directly *into* the cylinder.

We can follow the course of the engine as it goes through its cycle by following the pressure changes inside the cylinder. A diagram showing the pressure at each stage is called an *indicator diagram.* Indicator diagrams were used by James Watt, but he kept them a trade secret; the French scientist Emile Clapeyron introduced them into the discussion of the Carnot cycle. Carnot indeed has a deep debt of gratitude to Clapeyron, for not

The Carnot engine *consists of a working gas confined to a cylinder which may be put in thermal contact with hot or cold reservoirs, or thermally insulated, at various stages of the cycle of operations. Each stage of the cycle is performed* quasistatically *(infinitely slowly), and in a manner which ensures that the maximum amount of work is extracted. There are no losses arising from turbulence, friction, and so on.*

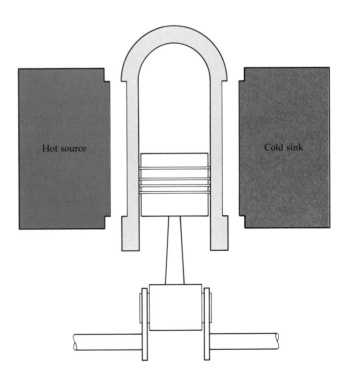

Hot source

Cold sink

James Watt (1736–1819).

Emile Clapeyron (1799–1864).

only did the latter refine his cycle, make a mathematical analysis of it, and portray it in terms of an indicator diagram, but it was Clapeyron's paper *"Mémoire sur la puissance motrice du feu"* (yet another variation on the theme), published in 1834, that kept Carnot's work alive and brought it to the attention of others, particularly of Kelvin.

In order to follow the engine through its cycle, we need to know some elementary properties of gases. The first is that as a given amount of gas is confined to ever-smaller volumes (as a piston is driven in), then its pressure increases. The magnitude of the increase depends on how the compression is carried out. If the gas is kept in contact with a heat sink (a *thermal reservoir*) of some kind (for instance, a water bath or a great block of iron), then its temperature remains the same, and the compression is called *isothermal*. Under these circumstances the rise in pressure follows one of the curves (*isotherms*) shown in the figure on the facing page. (These isotherms are mathematically *hyperbolas*, $p \propto 1/V$, as was established by Robert Boyle in the mid-seventeenth century; their precise form, however, is immaterial to this discussion.) Alternatively, the gas may be thermally insulated (the cylinder wrapped in insulating material). Under these circumstances no heat may leave or enter the gas, and the compression is called *adiabatic*. The experimental observation is that during an adiabatic compression the temperature of a gas rises. (We shall see the atomic reason for

The relation between the pressure and the volume of a gas depends on the conditions under which the expansion or compression takes place. If the temperature is held constant, the relation is expressed by Boyle's law *that the pressure is inversely proportional to the volume: this gives rise to the* isotherms *in the illustration. On the other hand, if the sample is thermally isolated, its temperature rises as it is compressed (and falls as it expands), and the dependence is as shown by the* adiabats.

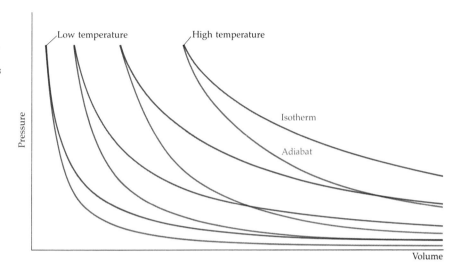

that later; for now, and throughout this chapter, we are keeping to the world of appearances and not delving into mechanisms.) The rise in temperature of the gas amplifies the rise in pressure that results from the confinement itself (because pressure increases with temperature); so, during an adiabatic compression, the pressure of a gas rises more sharply than during an isothermal compression (as is also illustrated above).

The increasing pressure of a gas as its volume is reduced isothermally, and the even sharper increase when the compression is adiabatic, are reversed when the gas expands. If the expansion is isothermal, then the pressure drops as the volume increases; if the expansion is adiabatic, then the pressure falls more sharply because the gas also cools. This is also shown in the figure above, which therefore summarizes most of the essential features of a gas.

The four steps of the Carnot cycle are illustrated at the top of the next page, and the behavior of the pressure of the confined gas is illustrated in the indicator diagram at the bottom of the next page. The horizontal axis in the latter represents the location of the piston, but because the cylinder is uniform it also represents the volume available to the gas; so from now on we shall interpret it as the volume.

The initial state of the engine is represented by *A* (in both illustrations). The hot source is in contact with the cylinder; so the gas is at the same temperature. The piston is in as far as it will go; so the volume is small. As a result of the high temperature and the confined volume, the pressure of the gas is high.

The Carnot cycle *consists of four stages:* A *to* B *is an isothermal expansion;* B *to* C *is an adiabatic expansion. Both steps produce work.* C *to* D *is an isothermal compression;* D *to* A *is an adiabatic compression. These two steps consume work. Each stage is traversed quasistatically.*

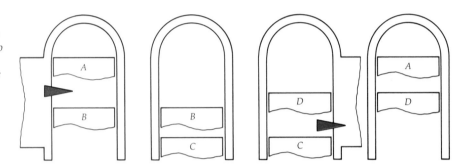

The first stage of the cycle is the expansion of the gas while the cylinder remains in contact with the hot source. The high-pressure gas pushes back the piston, and so the crank rotates. This is a power stroke of the engine. This step is isothermal (all at the same temperature); so, in order to overcome the tendency of the gas to cool as it expands, energy must flood in from the hot source. Therefore, not only is this the power stroke of the engine, it is also the step that sucks in energy—absorbs heat—from the hot source.

The designer of this engine could allow the crankshaft to continue to rotate, and so to return the piston to its initial position. Then, if the compression is also isothermal, the gas will be restored to its initial state. This would certainly fulfil one criterion: the cycle would be complete, the gas restored to its initial state, and the engine ready to go on again. We shall call this the *Atkins cycle*. Such a cycle is plainly useless, for in order to push the piston back to its starting position, exactly the same work needs to be

The indicator diagram *for the Carnot cycle.* AB *and* CD *are isotherms (from the figure on page 17), and* BC *and* DA *are adiabats (from the same illustration). The work produced during the cycle is proportional to the shaded area.*

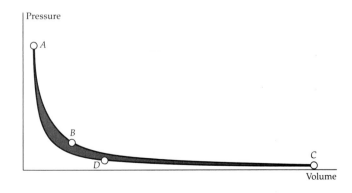

done by the external, initially hopeful, but now disappointed user as had been obtained from the engine in the power stroke! This is illustrated in the figure below, where the rotating crankshaft takes the state of the gas from A to B and back to A for ever, which is admirable, but the work generated in the first stage is reabsorbed in the second, which is not. Note, incidentally, that the area enclosed by the two lines that represent the useless Atkins cycle is zero: it is a standard result of elementary thermodynamics that the overall work produced during a complete cycle is given by the area it bounds on an indicator diagram. Here there is zero area between the two coincident curves: so zero work is produced overall.

The indicator diagram for the Atkins cycle. The two steps are isothermal, and occur at the same temperature: the cycle is useless, because no work is produced overall.

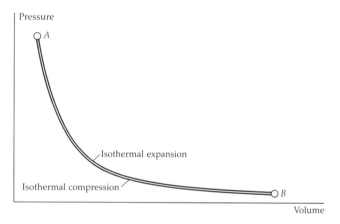

In order to make the cycle useful, we have to arrange matters so that not all the work of the power stroke is lost in restoring the gas to its initial pressure, temperature, and volume. We need a way to reduce the pressure of the gas inside the cylinder, so that during the compression stage less work has to be done to drive the piston back in. One way of reducing the pressure of a gas is to lower its temperature. That we can do by including in the cycle a stage of adiabatic expansion; we have seen that such an expansion lowers the temperature.

The essential step in the Carnot cycle is therefore to break the thermal contact with the hot reservoir before the piston is fully withdrawn, at B in the two figures opposite. The crank continues to turn, and the gas continues to expand. But now it does so adiabatically, and so both its pressure and its temperature fall. This takes its state to C. The stage from B to C is still a power stroke, but now we are cashing in the energy stored in the gas, for it can no longer draw on the supply from the hot source.

At this point we have to begin to restore the gas to its initial condition. The first restoration step involves pushing in the piston—doing work—and reducing the volume toward its initial value. This stage (from C to D) is performed with the gas in contact with the cold sink, to ensure that the pressure remains as low as possible and therefore that the work of compression is least. As the piston moves in, the gas tends to heat, but the thermal contact with the cold sink ensures that it remains at the same low temperature, for it can dump its extra energy into the sink.

This compression takes us to D. Now the volume of the confined gas is almost what it was initially, but its temperature is low. Therefore, before the crank has fully turned, we break the contact with the cold reservoir and allow the work of adiabatic compression to raise the temperature of the gas. If the timing is right, then the final surge of the piston not only compresses the gas to its initial volume, but also heats it to its initial temperature. The cycle is complete.

Not only is the cycle complete, and the engine back exactly in its initial condition and ready to go through another, but work has been produced. As we set out to achieve, more work was produced in the power strokes than was absorbed in the restoration stages, because the compression work has been done against a lower pressure. This is reflected in the shape of the indicator diagram at the bottom of page 18: it now encloses nonzero area, and the work done by the engine overall is also nonzero.

But there is an exceedingly important point. Notice the importance of the cold sink. Without the cold sink we have the primitive and useless Atkins cycle, a sequence that is cyclic but workless. As soon as we allow energy to be discarded into a cold sink, the lower line of the indicator diagram drops away from the upper, and the area that the cycle encloses becomes nonzero: the sequence is still cyclic, but now it is *useful.* However, the price we have to pay in order to generate work from the heat absorbed from the hot source is to throw some of that heat away. This captures the essence of Carnot's view that a heat engine is an energy mill (although we have discarded the conservation of caloric): *energy* drops from the hot source to the cold sink, and is conserved; but because we have set up this flow from hot to cold, we are able to draw only *some* energy off as work; so not all the energy drops into the cold. The cold sink appears to be essential, for only if it is available can we set up the energy fall, and draw off some as work.

Now we generalize. The Carnot cycle is only one way to extract work from heat. Nevertheless, it is the experience of everyone who has studied engines that, as in the Carnot cycle, *in every engine there has to be a cold sink,* and that at some stage of the cycle energy must be discarded into it. That little mouse of experience is nothing other than the Second Law of thermodynamics.

Thus the Second Law moves onto the stage. I have allowed it to creep in, because that emphasizes the extraordinary nature of thermodynamics. All the law seems to be saying is that heat cannot be completely converted into work in a cyclic engine: some has to be discarded into a cold sink. That is, we appear to have identified a fundamental tax: Nature accepts the equivalence of heat and work, but demands a contribution whenever heat is converted into work.

Note the dissymmetry. Nature does not tax the conversion of work into heat: we may fritter away our hard-won work by friction, and do so completely. It is only heat that cannot be so converted. Heat is taxed; not work.

The web of events is beginning to form. Bouncing balls come to rest; hot objects cool; and now we have recognized a dissymmetry between heat and work. The domain of the Second Law must now begin to spread outward from the steam engine and to claim its own. By the end of the book, we shall see that it will have claimed life itself.

2 THE SIGNPOST OF CHANGE

This is where we begin to define and refine corruption. So far we have seen that the immediate successors of Carnot were able to disentangle a rule about the quantity of energy from a rule about the direction of its conversion. Energy displaced heat as the eternally conserved; heat and work, hitherto regarded as equivalent, were shown to be dissymmetric. But these are bald, imprecise, and incomplete remarks: we must now sharpen them and put ourselves in a position to explore their ramifications. This we shall do in two stages. First, briefly, we shall refine the notions of heat and work, which so far we have regarded as "obvious" quantities. Then, with the precision such refinement will bring to the discussion, we shall start our main business, the refinement of the statement of the Second Law. With that refinement will come power and, as often happens, corruption too. We shall see that the domain of the Second Law is corruption and decay, and we shall see what extraordinarily wonderful things take place when quality gives way to chaos.

The Nature of Heat and Work

Central to our discussion so far, and for the next couple of chapters, are the concepts of *heat* and *work*. Perhaps the most important contribution of nineteenth-century thermodynamics to our comprehension of their nature has been the discovery that they are names of *methods*, not names of *things*. The early nineteenth-century view was that heat was a thing, the imponderable fluid "caloric"; but now we know that there is no such "thing" as heat. You cannot isolate heat in a bottle or pour it from one block of metal to another. The same is true of work: that too is not a thing; it can be neither stored nor poured.

Both heat and work are terms relating to the transfer of energy. *To heat* an object means to transfer energy to it in a special way (making use of a temperature difference between the hot and the heated). *To cool* an object is the negative of heating it: energy is transferred out of the object under the influence of a difference in temperature between the cold and the cooled. It is most important to realize, and to remember throughout the following pages (and maybe beyond), that *heat is not a form of energy: it is the name of a method for transferring energy.*

The same is true of work. Work is what you do when you need to change the energy of an object by a means that does not involve a temperature difference. Thus, lifting a weight from the floor and moving a truck to the top of a hill involve work. Like heat, *work is not a form of energy: it is the name of a method for transferring energy.*

All that having been established, we are going to return to informality again. In chapter 1 we said things like "heat was converted into work." If we were to speak precisely, we would have to say "energy was transferred from a source by heating and then transferred by doing work." But such precision would sink this account under a mass of verbiage; so we shall use the natural English way of talking about heat and work, and use expressions such as "heat flows into the system." But whenever we do, we shall always add in a whisper, "but we know what we really mean."

The Seeds of Change

Now we refine the Second Law into a constructive tool. So far it has crept mouselike into the discussion as a not particularly impressive commentary on some not particularly interesting experience with engines. Cold sinks, we have seen, are necessary when we seek to convert heat into work. The formal restatement of this item of experience is known as the *Kelvin statement* of the Second Law:

> **Second Law:** No process is possible in which the *sole result* is the absorption of heat from a reservoir and its *complete conversion* into work.

The most important point to pick out of this statement of the Second Law is the dissymmetry of Nature that we have already mentioned. It states that *it is impossible to convert heat completely into work* (see figure on left); it says nothing about the complete conversion of work into heat. Indeed, as far as we know, there is no constraint on the latter process: work may be completely converted into heat without there being any other discernable change. For example, frictional effects may dissipate the work being done by an engine, as when a brake is applied to a wheel. *All* the energy being transferred into the outside world by the engine may be dissipated in this way. Here, then, is Nature's fundamental dissymmetry; for although work and heat are equivalent in the sense that each is a manner of transferring energy, they are *not* equivalent in the manner in which they may interchange. We shall see that the world of events is the manifestation of the dissymmetry expressed by the Second Law.

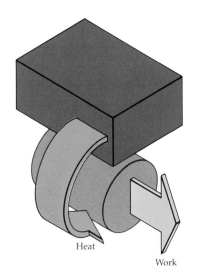

Heat

Work

The Kelvin statement of the Second Law denies the possibility of converting a given quantity of heat completely into work without other changes occurring elsewhere.

The Kelvin statement should not be construed too broadly. It denies the existence of processes in which heat is extracted from a source and converted completely into work, there being no other change in the Universe. It does not deny that heat can be completely converted into work when other changes are allowed to take place too. Thus cannons can fire cannonballs: the heat generated by the combustion of the charge is turned completely into the work of lifting the ball; however, cannons are literally one-shot processes, and the state of the system is quite different after the conversion (for instance, the volume of the gas that propelled the ball from the cannon remains large, and is not recompressed; cannons are not cycles).

One delight of thermodynamics is the way in which quite unrelated remarks turn out to be equivalent. This is the way the subject creeps over the landscape of events and digests them. Now the mouse can begin to grow and claim its own.

As an example of this process of incorporation, which allows the Second Law to spread away from the steam engine, we shall set in apparent opposition to the Kelvin statement of the Second Law the rival formulation devised by Clausius:

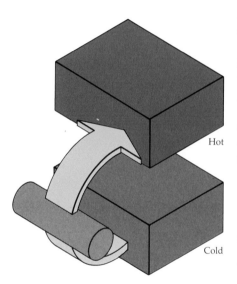

The Clausius statement of the Second Law denies the possibility of heat flowing spontaneously from a cold body to one that is hotter.

> **Second Law:** No process is possible in which the *sole result* is the transfer of energy from a cooler to a hotter body.

First, note that the Clausius statement can stand on its own as a summary of experience: so far as we know, no one has ever observed energy to transfer spontaneously (that is, without external intervention) from a cool body to a hot body (see figure on left). The laws of thermodynamics ignore, of course, the sporadic reports of purported miracles, and its proven predictive power is a retrospective argument against their occurrence. The fact that we need to construct elaborate devices to bring about refrigeration and air conditioning, and must run them by using electric power, is a practical manifestation of the validity of the Clausius statement of the Second Law: for although heat will not spontaneously flow to a hotter body, we can *cause* it flow in an unnatural direction if we allow changes to take place elsewhere in the Universe. In particular, a refrigerator operates at the expense of a burning lump of coal, a stream of falling water, or an exploding nucleus elsewhere. The Second Law specifies the *unnatural*, but does not forbid us to bring about the unnatural by means of a *natural* change elsewhere.

Second, the Clausius statement, like the Kelvin statement, identifies a fundamental dissymmetry of Nature, but ostensibly a different dissymmetry. In the Kelvin statement the dissymmetry is that between work and heat; in the Clausius statement there is no overt mention of work. The Clausius statement implies a dissymmetry in the direction of natural change: energy may flow spontaneously down the slope of temperature, not up. The twin dissymmetries are the anvils on which we shall forge the description of all natural change.

But there cannot be *two* Second Laws of thermodynamics: if the twin dissymmetries of Nature are both to survive, they must be the outcome of a *single* Second Law or at least one that should be expressed more richly than either the Kelvin or the Clausius statement alone. In fact, the two statements, although apparently different, are logically equivalent: there is indeed only one Second Law, and it may be expressed as either statement alone. The twin dissymmetries, and the anvils, are really one.

In order to show that the two statements are equivalent, we use the logical device of demonstrating that the Kelvin statement implies the Clausius statement, and that the Clausius statement implies the Kelvin. Actually, in the slippery way that logicians have, what we shall do is exactly the opposite: we shall show that if we can disprove the Kelvin statement, then the falsity of the Clausius statement is implied, and if we can disprove the Clausius, then farewell Kelvin too. If the death of either one implies the death of the other, then the statements are equivalent.

For our purposes, we bring on the family Rogue: Jack Rogue, the purveyor of anti-Kelvin devices, and Jill Rogue, whose line consists of anti-Clausius devices. First Jack will present his wares.

We take Jack's device, which he claims is an engine that contravenes Kelvin's experience, and can convert heat entirely into work and produce no change elsewhere, and we connect it between a hot source and a cold sink (see figure on facing page). We also connect it to another (conventional) engine, which will be run as a refrigerator and used to pump energy from the same cold sink to the same hot source. According to Jack, all the heat drawn from the hot source is converted into work. Suppose, then, that we run the engine long enough to remove 100 joules of energy* as heat, in which case, according to Jack, 100 joules of work are produced by his excellent machine. If that is so, then our other engine uses that 100 joules, and with it can transfer some energy from the cold sink to the hot source; the total energy it dumps as heat into that source is the sum of whatever it draws from the cold sink plus the 100 joules of energy that Jack's engine supplies. This must be so in order to accord with the First Law (which both Jack and Jill accept). These flows of energy are illustrated in the figure on the facing page. The overall effect, therefore, is to transfer

---

* The units for expressing quantities of energy, whether they are simply stored or are being shipped as heat or as work, are explained in Appendix 1. We shall use *joules.*

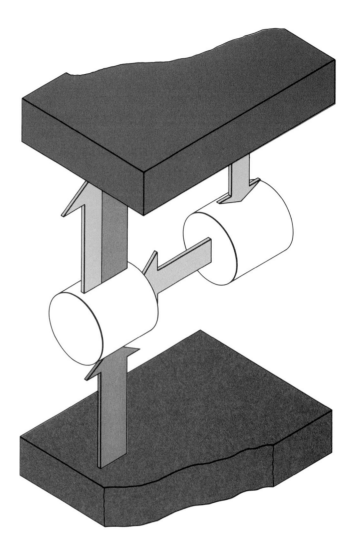

The argument to show that a failure of the Kelvin statement implies a failure of the Clausius statement involves connecting an ordinary engine between two reservoirs and driving it with an anti-Kelvin device. The net effect of the flows of energy shown here is to transport heat spontaneously from the cold to the hot reservoir, contradicting Clausius.

heat from cold to hot, there being no other change. Thus Jack's device pleases Jill.

Happy Jill now shows her device, which, she claims, spontaneously pumps heat from a cold sink to a hot source and leaves no change elsewhere. As was done with Jack's, Jill's device is connected between a hot source and a cold sink, and another engine is also connected between the two (see figure on next page). Jill runs her device, which pumps 100 joules of energy from cold to hot, and does so without any interference from outside, thus denying Clausius's experience of life. The other engine is arranged to run, and to dump 100 joules of energy into the cold sink, providing the balance of whatever it draws from the hot source as work.

In order to show that the failure of the Clausius statement implies a failure of the Kelvin statement, an ordinary engine is connected between hot and cold reservoirs which are also joined by an anti-Clausius device. The flows of energy are shown in the illustration, and the net effect is for a quantity of heat to be converted fully into work with no other change, contradicting Kelvin.

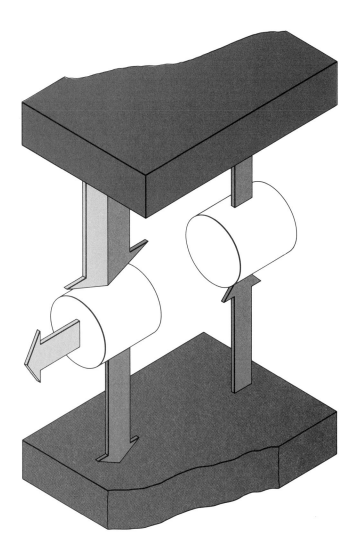

The flow of energy is shown above. Clearly, there has been no net change in the energy of the cold sink, and the overall outcome is for heat from the hot source to have been converted fully into work, with no change elsewhere, which pleases Jack.

Thus Jack and Jill are excellently matched: successful Jack, then successful Jill; successful Jill, then successful Jack. In other words, if Kelvin is false, then so is Clausius; and if Clausius is false, then so is Kelvin. Hence the Kelvin and Clausius statements are equivalent statements of experience: they are two faces of a single Second Law.

Toward Corruption

The progress of science is marked by the transformation of the qualitative into the quantitative. In this way not only do notions become turned into theories and lay themselves open to precise investigation, but the logical development of the notion becomes, in a sense, automated. Once a notion has been assembled mathematically, then its implications can be teased out in a rational, systematic way. Now, we have promised that this account of the Second Law will be nonmathematical, but that does not mean we cannot introduce a quantitative concept. Indeed, we have already met several, temperature and energy among them. Now is the time to do the same thing for spontaneity.

The idea behind the next move can be described as follows. The Zeroth Law of thermodynamics refers to the *thermal equilibrium* between objects ("objects," the things at the center of our attention, are normally referred to as *systems* in thermodynamics, and we shall use that term from now on). Thermal equilibrium exists when system *A* is put in thermal contact with system *B*, but no net flow of energy occurs. In order to express this condition, we need to introduce the idea of the *temperature* of a system, which we define as meaning that if *A* and *B* happen to have the same temperature, then we know without further ado that they are in thermal equilibrium with each other. That is, the Zeroth Law gives us a reason to introduce a "new" property of a system, so that we can easily decide whether or not that system would be in thermal equilibrium with any other system if they were in contact.

The First Law gives us a reason to carry out a similar procedure, but now one that leads to the idea of "energy." We may be interested in what states a system can reach if we heat it or do work on it. We can assess whether a particular state is accessible from the starting condition by introducing the concept of *energy*. If the new state differs in energy from the initial state by an amount that is different from the quantity of work or heating that we are doing, then we know at once, from the First Law, that that state cannot be reached: we have to do more or less work, or more or less heating, in order to bring the energy up to the appropriate value. The energy of a system is therefore a property we can use for deciding whether a particular state is accessible (see figure on next page).

This suggests that there may be a property of systems that could be introduced to accommodate what the Second Law is telling us. Such a property would tell us, essentially at a glance, not whether one state of the system is accessible from the other (that is the job of the energy acting through the First Law), but whether it is *spontaneously* accessible. That is, there ought to be a property that can act as the signpost of natural, sponta-

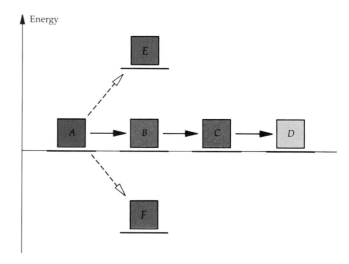

An isolated system may in principle change its state to any other of the same energy (the four colored boxes in the horizontal row), but the First Law forbids it to change to states of different energy (the brown-tinted boxes).

neous change, change that may occur without the need for our technology to intrude into the system in order to drive it.

There is such a property. It is the *entropy* of the system, perhaps the most famous and awe-inspiring thermodynamic property of all. Awe-inspiring it may be: but the awe should not be misplaced. The awe for entropy should be reserved for its power, not for its difficulty. The fact that in everyday discourse "entropy" is a word far less common than "energy" admittedly makes it less familiar, but that does not mean that it stands for a more difficult concept. In fact, I shall argue (and in the next chapter hope to demonstrate) that the entropy of a system is a simpler property to grasp than its energy! The exposure of the simplicity of entropy, however, has to await our encounter with atoms. Entropy is difficult only when we remain on the surface of appearances, as we do now.

Entropy

We are now going to build a working definition of entropy, using the information we already have at our disposal. The First Law instructs us to think about the energy of a system that is free from all external influences; that is, the constancy of energy refers to the energy of an *isolated system*, a system into which we cannot penetrate with heat or with work, and which for brevity we shall refer to as the *universe* (see figure on facing page). Similarly, the entropy we define will also refer to an isolated system, which we shall call the universe. Such names reflect the hubris of thermodynamics: later we shall see to what extent the "universe" is truly the Universe.

In thermodynamics we focus attention on a region called the system. *Around it are the surroundings. Together the two constitute the* universe. *In practice, the universe may be only a tiny fragment of the Universe itself, such as the interior of a thermally insulated, closed container, or a water bath maintained at constant temperature.*

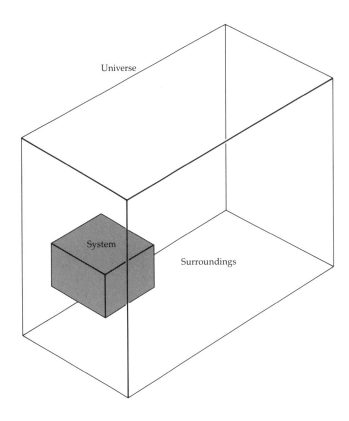

Suppose there are two states of the universe; for instance, in one a block of metal is hot, and in the other it is cold (see top figure on next page). Then the First Law tells us that the second state can be reached from the first only if the total energy of the universe is the same for each. The Second Law examines not the label specifying the energy of the universe, but another label that specifies the entropy. We shall define the entropy so that if it is *greater* in state *B* than in state *A,* then state *B may* be reached *spontaneously* from state *A* (see lower figure on next page). On the other hand, even though the energy of states *A* and *B* may be the same, if the entropy of state *B* is less than the entropy of state *A,* then state *B* cannot be reached spontaneously: in order to attain it, we would have to unzip the insulation of the universe, reach in with some technology (such as a refrigerator), and *drive* the universe from state *A* to state *B* (at the expense of a change in our larger Universe).

We have to construct a definition of entropy in such a way that in any universe entropy increases for natural changes, and decreases for changes that are unnatural and have to be contrived. Furthermore, we want to

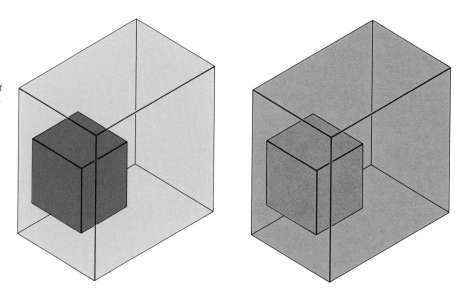

An isolated system (a universe) containing a hot block of metal is in a different state from one containing a similar but cold block, even if the total energy is the same in each. There must be a property other than total energy that determines the direction of spontaneous change will be hot $\longrightarrow$ cold *rather than the reverse.*

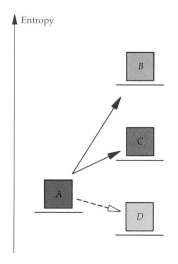

The states A, B, C, *and* D *in the illustration on page 30 have the same energy, but different entropies. The changes* A *to* B *and* A *to* C *may occur spontaneously, because each involves an increase of entropy; the change from* A *to* D *does not occur spontaneously, because it would require the entropy of the universe to drop. The universe always falls* upward *in entropy.*

define it so that we capture the Clausius and Kelvin statements of the Second Law, and arrive at a way of expressing them both simultaneously in the following single statement:

Second Law: Natural processes are accompanied by an increase in the entropy of the universe.

This is sometimes referred to not as the Second Law (which is properly a report on direct experience), but as the *entropy principle*, for it depends on a specification of the property "entropy," which is not a part of direct experience. (Similarly, the statement "energy is conserved" is also more correctly referred to as the *energy principle*, for the First Law itself is also a commentary on direct experience of the changes that work can bring about, whereas the more succinct statement depends on a specification of what is meant by "energy.")

The Kelvin statement is reproduced by the entropy principle if we define the entropy of a system in such a way that entropy increases when the system is heated, but remains the same when work is done. By implication, when a system is cooled its entropy decreases. Then Jack's engine is discounted by the Second Law, because heat is taken from a hot source (so that its entropy declines), and work is done on the surroundings (with the result that the entropy of the surroundings remains the same), as shown in the top figure on the facing page, and so overall the entropy of the little universe that contains his engine and its surroundings *decreases*; hence his engine is unnatural.

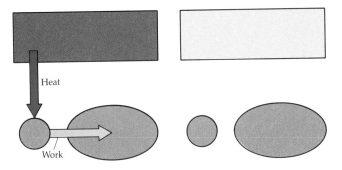

The shades of blue denote the entropies of the stored energy. When heat is withdrawn by the anti-Kelvin device, the entropy of the hot reservoir falls, but the quasistatic work does not produce entropy elsewhere. Overall, therefore, the entropy of the universe declines, which is against experience.

In order for us to discount Jill's device, the definition of entropy must depend on the temperature. We can capture her (and Clausius) if we suppose that the higher the temperature at which heat enters a system, the *smaller* the resulting change of entropy. In her anti-Clausius device, heat leaves the cold system, and the same quantity is dumped into the hot. Since the temperature of the cold reservoir is lower than that of the hot, the reduction of its entropy (see below) is greater than the increase of the entropy of the hot reservoir; so overall Jill's device reduces the entropy of the universe, and it is therefore unnatural.

Now the net is beginning to close in on natural change. We have succeeded in capturing Jack and Jill jointly on a single hook, just as we have claimed that the entropy principle captures the two statements of the Second Law. From now on we should be able to discuss *all* natural change in terms of the entropy.

Yet we are still hovering on the brink of actually defining entropy! Now is the time to take the plunge. We have seen that entropy increases when a

As in the illustration above, the shade of blue denotes the entropy. When heat is withdrawn from the cold reservoir, its entropy drops; when the same quantity of heat enters the hot reservoir, its entropy barely changes. Overall, therefore, the entropy of the universe declines, which is also against experience.

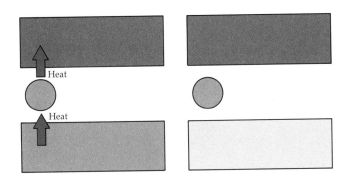

system is heated; we have seen that the increase is greater the lower the temperature. The simplest definition would therefore appear to be:

Change in entropy = (Heat supplied)/Temperature.

Happily, with care, this definition works.

First, let us make sure this definition captures what we have already done. If energy is supplied by heating a system, then *Heat supplied* is positive, and so the change of entropy is also positive (that is, the entropy increases). Conversely, if the energy leaks away as heat to the surroundings, *Heat supplied* is negative, and so the entropy decreases. If energy is supplied as work and not as heat, then *Heat supplied* is zero, and the entropy remains the same. If the heating takes place at high temperature, then *Temperature* has a large value; so for a given amount of heating, the change of entropy is small. If the heating takes place at low temperatures, then *Temperature* has a small value; so for the same amount of heating, the change of entropy is large. All this is exactly what we want.

Now for the care in the use of the definition. The temperature must be constant throughout the transfer of the energy as heat (otherwise the formula would be meaningless). Generally a system gets hotter (that is, its temperature will rise) as heating proceeds. However, if the system is extremely large (for example, if it is connected to all the rest of the actual Universe), then however much heat flows in, its temperature remains the same. Such a component of the universe is called a thermal *reservoir*. Therefore we can safely use the definition of the change of entropy *only* for a reservoir. That is the first limitation (it may seem extreme, but we shall spread the boundaries of the definition in a moment).

A second point concerns the manner in which energy is transferred. Suppose we allow an engine to do some work on its surroundings. Unless we are exceptionally careful, the raising of the weight, the turning of the crank, or whatever, will give rise to turbulence and vibration, which will fritter energy away by friction and in effect heat the surroundings. In that case we would expect the transfer of energy as work also to contribute to the change in entropy. In order to eliminate this from the definition (but once again only in order to clarify the definition, not to eliminate dissipative processes from the discussion), we must specify how the energy is to be transferred. The energy must be transferred without generating turbulence, vortices, and eddies. That is, it has to be done infinitely carefully: pistons must be allowed to emerge infinitely slowly, and energy must be allowed to seep down a temperature gradient infinitely slowly. Such processes are then called *quasistatic:* they are the limits of processes carried out with ever-increasing care.

Measuring the Entropy

We have a definition of entropy, but the definition does not seem to give the concept much body. Although we regard properties such as temperature and energy to be "tangible" (but we do so merely because they are familiar), the idea of entropy as (*Heat supplied*)/*Temperature* seems remote from experience. So it is, and so it will remain until the next chapter, where we shall add flesh by considering how to interpret the concept in terms of the behavior of atoms.

But is temperature *really* so familiar, and entropy so remote? We think of a liter of hot water and a liter of cold water as having different temperatures. *In fact,* they also have different entropies, and the "hot" water has both a higher entropy and a higher temperature than the cold water. The fact that hot water added to cold results in tepid water is a consequence of the change of entropy. Should we think then of "hotness" as denoting high temperature or as denoting high entropy? With which concept are *really* familiar?

Temperature seems familiar because we can measure it: we feel at home with pointer readings, and often mistake the reading for the concept. Take time, for instance: the pointer readings are an everyday commonplace, but the *essence* of time is much deeper. So it is with temperature; although it seems familiar, the nature of temperature is a far more subtle concept. The difficulty with accepting entropy is that we are not familiar with instruments that measure it, and consequently we are not familiar with their pointer readings. The *essence* of entropy, when we get to it, is certainly no more difficult, and may be simpler, than the essence of temperature. What we need, therefore, in order to break down the barrier between us and entropy, is an entropy meter.

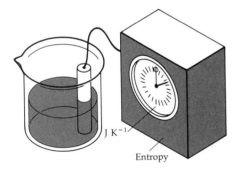

An entropy meter consists of a probe in the sample and a pointer giving a reading on a dial, exactly like a thermometer.

The figure to the left shows an entropy meter; the figure on the next page indicates the sort of mechanism that we might find inside it: it is basically a thermometer attached to a microprocessor. The readings can be taken from the digital display.

Suppose we want to measure the entropy change when a lump of iron is heated. All we need do is attach the entropy meter to the lump, and start heating: the microprocessor monitors the temperature indicated by the thermometer, and converts it directly into an entropy change. What calculations it does we shall come to in a moment. The care we have to exercise is to do the heating extremely slowly, so that we do not create hot spots and get a distorted reading: the heating must be quasistatic.

The microprocessor is programmed as follows. First, it has to work out, from the rise in temperature caused by the heating, the quantity of energy that has been transferred to the lump from the heater. That is a fairly

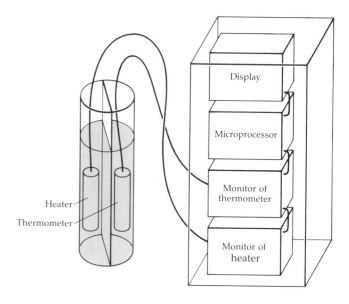

The interior of the entropy meter is more complicated than that of a simple mercury thermometer. The probe consists of a heater (whose output is monitored by the rest of the meter) and a thermometer (which is also monitored). The microprocessor is programmed to do a calculation based on how the temperature of the sample depends on the heat supplied by the heater. The output shown on the dial is the entropy change of the sample between the starting and finishing temperatures.

straightforward calculation once we know the *heat capacity* (the *specific heat*) of the sample, because the temperature rise is directly proportional to the heat supplied:

Temperature rise = (Proportionality coefficient) × (Heat supplied),

the coefficient being related to the heat capacity. (We could always measure the heat capacity in a separate experiment, with the same apparatus, but with a different program in the microprocessor.) The heater supplies only a trickle of energy to the sample, and the microprocessor evaluates *(Heat supplied)/Temperature,* and stores the result. If only a little heat is supplied, the temperature will hardly rise, and so the entropy formula is very accurate. However, since the sample is not an infinite reservoir, the temperature does rise a little, and the next trickle of heat takes place at a slightly higher temperature. The microprocessor therefore has to evaluate the next trickle of *(Heat supplied)/Temperature* at a marginally (in the limit, infinitesimally) higher temperature. It adds the result to the previous value (see the figure on the facing page).

   The procedure continues: the thermometer records, the microprocessor goes on dividing and adding, and the heating continues until at long last (in a perfect experiment, at the other end of eternity) the temperature has risen to the final value. The microprocessor then displays the accumulated sum of all the little values of *(Heat supplied)/Temperature* as the change in entropy of the lump.

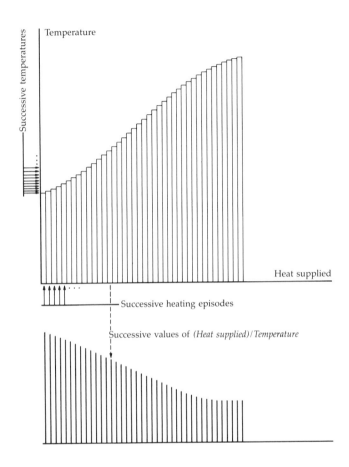

The entropy meter works by squirting tiny quantities of heat into the sample, and monitoring the temperature. It then evaluates (Heat supplied)/Temperature, *and stores the result. Next it monitors the new temperature, and squirts in some more heat, and repeats the calculation. This is repeated until the final temperature has been reached. In a real-life measurement, the* heat capacity *of the sample is measured over the temperature range, and the entropy change is calculated from that (see Appendix 2).*

That is as far as we need go for now. What I want to establish here is not so much the details of how the entropy change is measured in any particular process, but the fact that it is a measurable quantity, exactly like the temperature, and, indeed, that it can be measured with a thermometer too!

The Dissipation of Quality

We can edge closer to complete understanding by reflecting on the implications of what this external view of entropy already reveals about the nature of the world. As a first step, we shall see how the introduction of entropy leads to a particularly important interpretation of the role of energy in events.

Suppose we have a certain amount of energy that we can draw from a hot source, and an engine to convert it into work. We know that the Second Law demands that we have a cold sink too; so we arrange for the engine to operate in the usual way. We can extract the appropriate quantity of work, and pay our tax to Nature by dumping a contribution of energy as heat into the cold sink. The energy we have dumped into the cold sink is then no longer available for doing work (unless we happen to have an even colder reservoir available). Therefore, in some sense, energy stored at a high temperature has a better "quality": high-quality energy is available for doing work; low-quality energy, corrupted energy, is less available for doing work.

A slightly different way of looking at the quality of energy is to think in terms of entropy. Suppose we withdraw a quantity of energy as heat from the hot source, and allow it to go directly to the cold sink (see the figure to the left). The entropy of the universe decreases by an amount $(Heat\ withdrawn)/Temperature_{HOT\ SOURCE}$, but also increases by an amount $(Heat\ dumped)/Temperature_{COLD\ SINK}$. The sum of the two contributions to the overall change in entropy is therefore positive (because the temperature of the hot source is higher than that of the cold sink). The energy of the universe is then less available for doing work (because when energy is stored at lower temperatures, still colder sinks are needed if it is to be converted into work). It is then, in our sense, lower in quality, and the entropy associated with the energy has increased. The *entropy*, therefore, *labels the manner in which the energy is stored*: if it is stored at a high temperature, then its entropy is relatively low, and its quality is high. On the other hand, if the same amount of energy is stored at a low temperature, then the entropy of that energy is high, and its quality is low.

Just as the increasing entropy of the universe is the signpost of natural change and corresponds to energy being stored at ever-lower temperatures, so we can say that *the natural direction of change is the one that causes the quality of energy to decline:* the natural processes of the world are manifestations of this corruption of quality.

This attitude toward energy and entropy, that entropy represents the manner in which energy is stored, is of great practical significance. The First Law establishes that the energy of a universe (and maybe of the Universe itself) is constant (perhaps constant at zero). Therefore, when we burn fossil fuels, such as coal, oil, and nuclei, we are not diminishing the supply of energy. In that sense, there can never be an energy crisis, for the energy of the world is forever the same. However, every time we burn a lump of coal or a drop of oil, and whenever a nucleus falls apart, we are increasing the entropy of the world (for all these are spontaneous processes). Put another way, every action diminishes the quality of the energy of the universe.

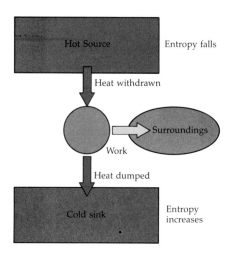

Some heat must be discarded into a cold sink in order for us to generate enough entropy to overcome the decline taking place in the hot reservoir.

As technological society ever more vigorously burns its resources, so the entropy of the universe inexorably increases, and the quality of the energy it stores concomitantly declines. We are not in the midst of an energy crisis: we are on the threshold of an entropy crisis. Modern civilization is living off the corruption of the stores of energy in the Universe. What we need to do is not to conserve energy, for Nature does that automatically, but to husband its quality. In other words, we have to find ways of furthering and maintaining our civilization with a lower production of entropy: the conservation of quality is the essence of the problem and our duty toward the future.

Thermodynamics, particularly the Second Law (we shall see the less than benign role of the Third in a moment), indicates the problems in this program of conservation, and also points to solutions. In order to see how this is so, we shall go back to the Carnot cycle, and apply what we have developed here to its operation.

Ceilings to Efficiency

In the first place, if the Carnot engine goes through its cycle, then the entropy change of its little world cannot be negative, for that would signify a nonspontaneous process, and useful engines do not have to be driven. Now, however, we are equipped to calculate the change in entropy, using the formula *Heat/Temperature*. In order to calculate however, we must assume that the engine is working perfectly, and that there are no losses of any kind: the cycle must be gone round quasistatically.

The engine itself returns to its initial condition (it is cyclic); so at the end of a cycle it has the same entropy as it had at the beginning. The work it does in the surroundings does not increase their entropy, because everything happens so carefully and slowly in the quasistatic operating regime. The only changes of entropy are in the hot source, the entropy of which *decreases* by an amount of magnitude

$$\text{(Heat supplied from hot source)}/T_{\text{HOT SOURCE}},$$

and in the cold sink, the entropy of which *increases* by an amount of magnitude

$$\text{(Heat supplied to cold sink)}/T_{\text{COLD SINK}}.$$

And, under quasistatic conditions, that is all. However, overall the change of entropy must not be negative. Therefore *the smallest value of the heat discarded into the cold sink must be large enough to increase the entropy there just enough to overcome the decrease in entropy in the hot source*. It is straightforward algebra to show that this minimum discarded energy is

Minimum heat discarded into the cold sink
$$= (\text{Heat supplied by hot source}) \times (T_{\text{COLD SINK}} / T_{\text{HOT SOURCE}}).$$

Here is our first major result of thermodynamics: we now know how to minimize the heat we throw away: we keep the cold sink as cold as possible, and the hot source as hot as possible. That is why modern power stations use superheated steam: cold sinks are hard to come by; so the most economical procedure is to use as hot a source as possible. That is, the designer aims to use the highest-quality energy.

But we can go on, and summon up our second major result. The work generated by the Carnot engine as it goes through its cycle must be equal to the difference between the heats supplied and discarded (this is a consequence of the First Law). The work is therefore equal to *Heat supplied minus Heat discarded* (see the preceding figure). We are now, however, in a position to express this difference in terms of the *Heat supplied* multiplied by a factor involving the two temperatures. The *efficiency* of the engine is the ratio of the work it generates to the heat it absorbs. It is now very simple to arrive at the result that the efficiency of a Carnot engine, working perfectly between a hot source and a cold sink, is

$$Efficiency = 1 - (T_{\text{COLD SINK}} / T_{\text{HOT SOURCE}}).$$

That is, the efficiency depends only on the temperatures and is independent of the working material in the engine, which could be air, mercury, steam, or whatever. Most modern power plants for electricity generation use steam at around 1,000 °F (800 K) and cold sinks at around 212 °F (373 K).* Their efficiency ceiling is therefore around 54 percent (but other losses reduce this efficiency to around 40 percent). Higher source temperatures could improve efficiencies, but bring other problems, because then materials begin to fail. For safety reasons, nuclear reactors operate at lower source temperatures (of about 600 °F, 620 K), which limits their theoretical efficiency to around 40 percent. Losses then reduce this figure to about 32 percent. Closer to home, an automobile engine operates with a briefly maintained input temperature of over 5,400 °F (around 3,300 K) and exhausts at around 2,100 °F (1,400 K), giving a theoretical ceiling of around 56 percent. However, actual automobile engines are designed to be light enough to be responsive and mobile, and therefore attain only about 25 percent efficiency.

---

* Scales of temperature are described in Appendix 1. K denotes *kelvin*, the graduation of the Kelvin scale of temperature (the one of fundamental significance, in contrast to the contrived scales of Celsius and Fahrenheit). In brief, a temperature in kelvins is obtained by adding 273 to the temperature in degrees Celsius.

The profound importance of the preceding result is that is puts an upper limit on the efficiency of engines: whatever clever mechanism is contrived, so long as the engineer is stuck with fixed temperatures for the source and the sink, the efficiency of the engine cannot exceed the Carnot value. The reason why should by now be clear (to the external observer). In order for heat to be converted to work spontaneously, there must be an overall increase in the entropy of the universe. When energy is withdrawn as heat from the hot source, there is a reduction in its entropy. Therefore, since the perfectly operating engine does not itself generate entropy, there must be entropy generated elsewhere. Hence, in order for the engine to operate, there must be a dump for at least a little heat: there must be a sink. Moreover, that sink must be a cold one, so that even a small quantity of heat supplied to it results in a large increase in entropy.

The temperature of the cold sink amplifies the effect of dumping the heat: the lower the temperature, the higher the magnification of the entropy. Consequently, the lower the temperature, the less heat we need to discard into it in order to achieve an overall positive entropy change in the universe during the cycle. Hence the efficiency of the conversion increases as the temperature of the cold source is lowered.

There appears to be a limit to the lowness of temperature. The conversion efficiency of heat to work cannot exceed unity, for otherwise the First Law would be contravened. Therefore the value of $Temperature_{\text{COLD SINK}}$ cannot be negative. Hence there appears to be a *natural limit to the lowness of temperature*, corresponding to $Temperature_{\text{COLD SINK}} = 0$. This is the *absolute zero of temperature*, the end of getting cold. At this infinite arctic, the conversion efficiency would be unity, for even the merest wisp of heat transferred to the sink would give an enormous positive entropy (because the temperature is in the denominator, so that $1/Temperature$ becomes infinitely large and magnifies everything infinitely). But can we attain that Nirvana?

A clue to the attainability of absolute zero can be obtained by considering the Carnot cycle with an ever-decreasing temperature of its cold sink. For a given quantity of heat to be absorbed from the hot source, the piston needs to travel out a definite distance from A to B in the figure on page 18, no matter what becomes of the energy later. The cooling step, the adiabatic expansion from C to D, then involves a greater expansion the lower the temperature we are aiming to reach. Some of the expansions are illustrated in the figure on the next page: we can see that the lower the temperature aimed at, the greater the size of the stroke. In order to approach very low temperatures, we need extremely large engines. In order to reduce the temperature to zero, we would need an infinitely large engine. Absolute zero appears to be unattainable.

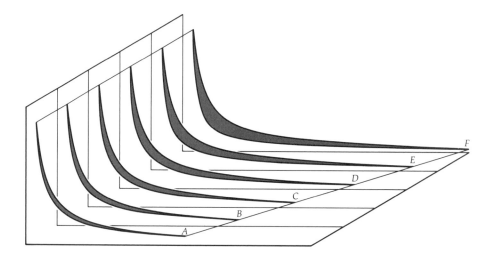

Carnot indicator diagrams for cycles with decreasing cold-sink temperatures (F is coldest) but constant heat input. The work output (shaded area) increases, and therefore so does the efficiency, but the stroke required becomes large.

The *Third Law* of thermodynamics generalizes this result. In a dejected kind of way it summarizes experience by the following remark:

Third Law: Absolute zero is unattainable in a finite number of steps.

This gives rise to the following sardonic summary of thermodynamics:

First Law: Heat can be converted into work.
Second Law: But completely only at absolute zero.
Third Law: And absolute zero is unattainable!

The End of the External

We have traveled a long way in this chapter. First, we drew together the skeins of experience summarized by the Kelvin and the Clausius statements of the Second Law, saw that they were equivalent, and exposed two faces of Nature's dissymmetry. We also saw that we could draw the two statements together by introducing a property of the system not readily discernable to the untutored eye, the entropy. We have seen that the entropy may be measured, and that it may be deployed to draw far-reaching conclusions about the nature of change. We have seen that the Universe is rolling uphill in entropy, and that it is thriving off the corruption of the quality of its energy.

Yet all this is superficial. We have been standing outside the world of events, but we have not yet discerned the deeper nature of change. Now is the time to descend into matter.

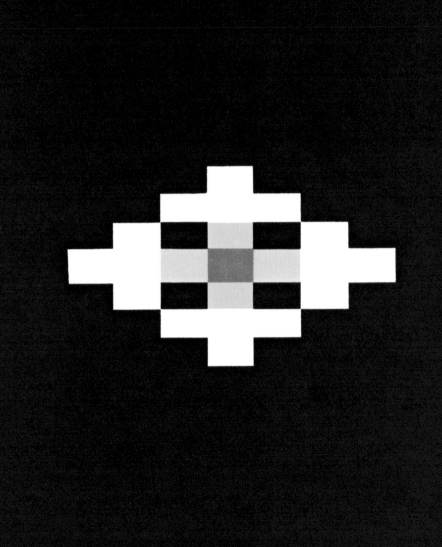

3 COLLAPSE INTO CHAOS

Matter consists of atoms. That is the first step away from the superficialities of experience. Of course, we could burrow even further beneath superficiality, and regard matter as consisting of more (but perhaps not limitlessly more) fundamental entities. Perhaps Kelvin was right in his suspicion that the most fundamental aspect of the world is its eternal, elusive, and perhaps zero energy. But although the onion of matter can be peeled beyond the atom, that is where we stop; for in thermodynamics we are concerned with the changes that occur under the gentle persuasion of heat, and under most of the conditions we encounter, the energy supplied as we heat a system is not great enough to break open its atoms. The gentleness of the domain of thermodynamics is why it was among the first of scientists' targets: only as increasingly energetic methods of exploration and destruction became available were other targets opened to inspection, and in turn, as the vigor of wars increased, so the internal structure of the atom, the nucleus, and the nucleons became a part of science. Heat, although it may burn and sear, is largely gentle to atoms.

The concept of the atom, although it originated with the Greeks, began to be convincing during the early nineteenth century, and came to full fruition in the early twentieth. As it grew, there developed the realization that although thermodynamics was an increasingly elegant, logical, and self-sufficient subject, it would remain incomplete until its relation to the atomic model of matter had been established. There was some opposition to this view; but support for it came from (among others) Clausius, who identified the nature of heat and work in atomic terms, and set alight the flame that Boltzmann soon was to shine on the world.

Although we have been speaking of *atoms*, in many of the applications of thermodynamics *molecules* also play an important role, as do *ions*, which are atoms or molecules that carry an electric charge. In order to cover all these species, we shall in general speak of *particles*.

Inside Energy

As a first step into matter we must refine our understanding of energy by recalling some elementary physics. In particular, we should recall that a particle may possess energy by virtue of its location and its motion. The former is called its *potential energy*; the latter is its *kinetic energy*.

A particle in the Earth's gravitational field has a potential energy that depends on its height: the higher it is, the greater its potential energy. Likewise, a spring has a potential energy that depends on its degree of extension or compression. Charged particles near each other have a potential energy by virtue of their electrostatic interaction. Atoms near each other have a potential energy by virtue of their interaction (largely the electrostatic interactions between their nuclei and their electrons).

A moving particle possesses kinetic energy: the faster it goes, the greater its kinetic energy. A stationary particle possesses no kinetic energy. A heavy particle moving quickly, like a cannonball (or, in more modern terms, a proton in an accelerator), possesses a great store of energy in its motion.

The most important property of the *total energy* of a particle (the sum of its potential and kinetic energies) is that it is constant in the absence of externally applied forces. This is the law of the *conservation of energy*, which moved to the center of the stage as the importance of energy as a unifying concept was recognized during the nineteenth century. It accounts for the motion of everyday particles like baseballs and boulders, and applies to particles the size of atoms (subject to some subtle restrictions of that great clarifier, the quantum theory). For instance, the law readily accounts for the motion of a pendulum: there is a periodic conversion from potential to kinetic energy as the bob swings from its high, stationary turning point, moves quickly (with high kinetic energy) through the region of lowest potential energy (at the lowest point of its swing), and then climbs more and more slowly to its next turning point. Potential and kinetic energy are equivalent, in the sense that one may readily be changed into the other; their sum, in an isolated object, remains the same.

Intrinsic to the soul of thermodynamics is the fact that it deals with vast numbers of particles. A typical yardstick to keep in mind is *Avogadro's number*. Its value is about 6×10^{23}, and it represents the number of atoms in 12 grams of carbon. (By coincidence, it is not far off the number of stars in all the galaxies in the visible Universe.) The idea to appreciate here is not the precise value of Avogadro's number, or the precise number of atoms in any given system, but the fact that the numbers of atoms involved in everyday samples of matter are truly enormous. It may seem surprising at first sight that science learned to deal with the properties of such enormous crowds of particles before it discovered how to deal with individual atoms. The reason lies at the core of thermodynamics: the thermodynamic properties of a system are *average* values over statistically large assemblies of particles. Just as it is easier to deal with average properties of human populations than with individuals, be they consumers, wearers, or wage-earners, so it is easier to deal with the average properties of assemblies of particles than with the individuals. Idiosyncracies (which the atomically aware thermodynamicist terms *fluctuations*) are ironed out and become relatively insignificant when the populations are large, and the population of particles in a typical sample is vastly greater than the population of people of any nation.

The *energy* of a thermodynamic system, such as the several Avogadro's numbers of water molecules in a glass of water, is the sum of the kinetic energies of all the particles and of their potential energies too. Hence, it should be plain that this total energy is constant (the essential content of simple versions of the First Law). However, in a many-particle thermodynamic system, a new aspect of the motion, one not open to a single particle on its own becomes available.

Consider the kinetic energy of the collection. If all the particles happen to be traveling in the same direction with the same speed, then the entire system is in flight, like a baseball (see below). The entire system behaves like a single, massive particle, and the ordinary laws of dynamics apply.

However, there is another sort of motion. Instead of all the particles moving uniformly, we can think of them as being chaotic: the *total* energy of the system may be the same as that of the ball in flight, but now there is no net motion, because all the directions and speeds of the atoms are jumbled up in chaos (see the figure on the next page). If we could follow any

The particles of a ball in flight are moving co-herently: they are all turned ON*.

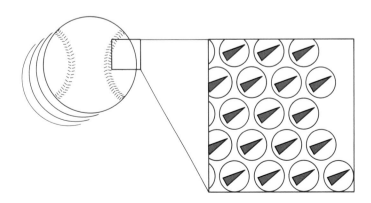

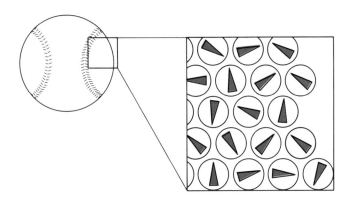

The same quantity of energy may be stored by a stationary but warm ball. Now the particles are moving incoherently: they are turned ON. The random, incoherent motion is called thermal motion.

individual particle, we would see it moving a tiny amount to the right, bouncing off its neighbor, moving to the left, bouncing again, and so on. The central feature is the lack of *correlation* between the motions of different particles: their motion is *incoherent*.

This random, chaotic, uncorrelated, incoherent motion is called *thermal motion*. Obviously, since it is meaningless to speak of the uncorrelated motion of a single particle, the concept of thermal motion cannot be applied to single particles. In other words, when we step from considering a single particle to considering systems of many particles, when the question of coherence becomes relevant, we are stepping out of simple dynamics into a new world of physics. This world is *thermo*dynamics. All the richness of the subject, the way that the steam engine can make the journey into life and account for the unfolding of a leaf, results from this enlargement of domain.

We have established that there are two modes of motion for the particles of a composite system: the motion may be coherent, when all the particles are in step, or the motion may be incoherent, when the particles are moving chaotically. We have also seen in our encounter with the First Law that there are two modes of transferring energy to a system, by doing work on it or by heating it. Now we can put the remarks together:

When we do *work* on a system, we are stimulating its particles with *coherent* motion; when the system is doing work on the surroundings, it is stimulating coherent motion.

When we *heat* a system, we are stimulating its particles with *incoherent* motion; when a system is heating its surroundings, it is stimulating incoherent motion.

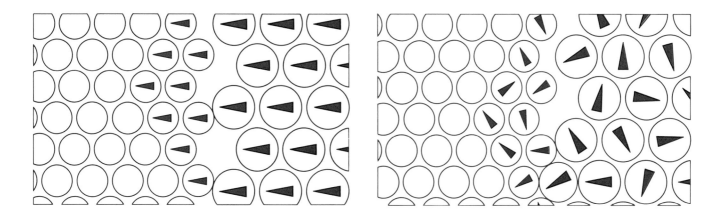

Work involves the transfer of energy by using the coherent *motion of particles in the surroundings (left illustration). The particles in the system will pick up the coherent motion, and might dissipate it into thermal motion. Heat is the transfer of energy by using the* incoherent *motion of particles in the surroundings (right illustration). They jostle the particles of the system into incoherent thermal motion.*

This distinction is illustrated above.

A couple of examples should make this clear. Suppose we want to change the energy of a 1-kilogram block of iron (a cube about 5 cm on each side). One way would be to lift it: lifting it through 1 meter increases its *potential* energy by about 10 joules (Appendix 1). What we have done is move all its atoms *coherently* through a displacement of 1 meter. Energy has been transferred to the block, and is now stored in the gravitational potential energy of all its atoms. Energy has been transferred by doing *work*.

Suppose instead that the block is hurled off in some horizontal direction. Now the *kinetic* energy of all its atoms has been increased and their motion is coherent. If they all move at 4.5 meters per second (about 10 m.p.h.), the block acquires 10 joules of energy. Energy has been transferred to the block, and is now stored in the kinetic energy of all its atoms. Once again, energy has been transferred by doing *work*.

Now suppose we expose the block to a flame, and raise its temperature. This increases its energy, but the block remains in its initial position and seems not to be moving. However, if the temperature is raised by only 0.03 °C, the transfer will correspond to 10 joules of energy, exactly as before. Now the energy is stored in the *thermal motion* of the atoms. It is still stored as their kinetic and potential energies (the *only* form of energy storage we ever need consider), but now the locations and motions of the atoms are incoherent, and there is no net displacement or motion of the block as a whole. Energy has been transferred to the block by stimulating the incoherent motion of its atoms. Energy has been transferred by *heating* the block.

The Mark I universe. *Each rectangle represents an atom; there are 1,600 in all.*

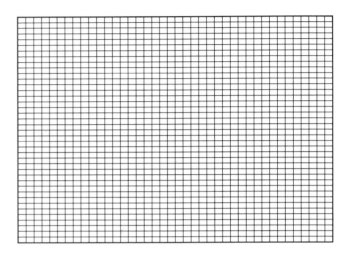

The Mark I universe. *Each rectangle represents an atom; there are 1,600 in all.*

Modeling the Universe

The Universe is quite a complicated place, but there are a lot of simple
things going on inside it. In much of the following discussion, it will prove
useful to focus on the essential features of the processes without getting
distracted by complications like dogs, opinions, and other trappings of
reality. Of course, we must always ensure that the simplification doesn't
destroy important details; so we shall swing frequently between the very
simple models of the Universe and the actual Universe, in which
simplicities are sometimes so cloaked in consequences that their true na-
tures are obscured. When we refer to a model of the Universe, we shall call
it a "universe."

We shall use two simple models of the Universe. The *Mark I universe*
consists of up to 1,600 atoms (see above).* Each atom may be energetically
unexcited, which we shall call OFF, or energetically excited, which we shall
call ON. In the illustrations we shall denote ON-ness by a red blob (see
figure on left). When several atoms are ON, we shall take their motions to

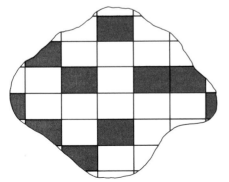

*An atom of the model universe that is storing
energy is said to be "ON": ON-ness is denoted
by red, OFF-ness by white. The ON atoms can
be thought of as vibrating incoherently, and as
having exactly the same energy as any other
atom that is ON.*

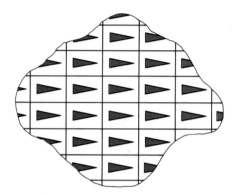

An atom of the model universe that is storing energy in a way that is correlated with the energy stored in other atoms (for instance, in uniform motion) is said to be ON*: ON*-ness is denoted by a red arrow. Once again, each atom can store a single characteristic quantity of energy.

be uncorrelated unless specified. That is, several red blobs in the region of the universe that represents some system means that the atoms are storing their energy as thermal motion. When we want to indicate that the motion of a group of atoms is coherent, we shall say that the atoms are ON* and denote them by red arrows in the direction of their motion (see figure on left). Another simplifying feature of the Mark I universe is that each atom can possess only a single characteristic energy when it is ON, and this quantity is the same for each atom. (We could think of the energy required to turn an atom ON as being 1 joule, although that would correspond to a cannonball of an atom. If a hydrogen atom is supplied with only 10^{-18} joules, it falls apart. The actual value will not be important for most of what follows; so we can adopt the simplest value in order to have something concrete, or just leave it unspecified if something more concrete is not needed.)

The *Mark II universe* is the same as the Mark I version, except that the number of atoms in it is infinite: we can still use the 1,600 blobs to show atoms, but now this represents only a minute fraction of the total universe (see below). This is the universe we use when we want to model the reservoirs we mentioned in Chapter 2: they are insatiable sinks and inexhaustible sources.

The *Mark III universe* (see figure on next page) we shall hold in reserve for now: it has more complicated entities, such as atoms that can possess various quantities of energy, atoms of different kinds, concatenations of atoms, and even people.

The Mark II universe *is like the Mark I version, but it has an indefinitely large number of identical atoms.*

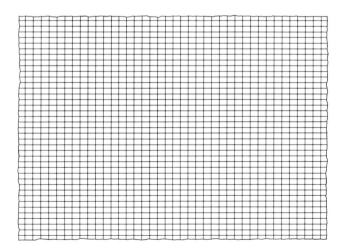

The Mark III universe *is much more compli-cated: it has all sorts of atoms strung together in complex patterns. Nevertheless, the underly-ing processes are no more complex than the ones possessed by the earlier Marks.*

Now we put the Mark I universe into operation and see what it reveals about the process of change. The only rule we shall impose is the conserva-tion of energy in the universe as a whole (so that the number of atoms ON remains constant). We shall allow any atom to hand on its ON-ness to a neighbor, or pick up energy, and so turn ON, from a neighbor that hap-pens to be ON already. (An atom cannot be more than ON or partially ON: one ON per atom at most; and either ON or OFF.)

In terms of an actual process in the Universe, we can think of the gray area in the figure on the facing page as representing one block of iron, and the untinted area as representing another. Then the state of being ON represents an atom that is vibrating vigorously around its average location, and an atom that is OFF represents an atom that is motionless. The hand-ing on of ON-ness then corresponds to an ON atom jostling a previously motionless neighbor, which breaks into vibration at the expense of the energy of the previously vibrating atom. This handing on of vibrational motion is a random process, and so ON-ness just wanders at random from atom to atom.

The central point about the behavior of the universe, and by extension of the Universe too, is that *properties arise from the minimum of rules*. The only

The initial state of a Mark I universe modeling two blocks of metal in contact. No atoms in the bigger (L-shaped) block are ON, and all the energy of the universe resides in the smaller block, which is hot. Using the formula we give later, we can calculate that the temperature of System 1 is 2.47 and that of System 2 is zero.

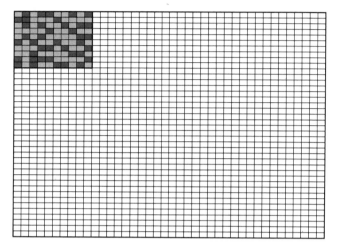

rule we are adopting is the conservation of ON-ness. We are allowing undirected and unconstrained mobility. Even with this light touch, the universe possesses properties. The same properties could be *contrived* by imposing rules, such as the rule that energy shall migrate from atom to atom in a specified manner. But such rules are plainly unnecessary, and the scientific razor cuts them out.

Suppose we have the arrangement of ON-ness in the universe as depicted above. This corresponds to a lot of energy stored in the thermal motion of System 1 (one block of iron), and none at all in System 2 (the other block). What will happen?

As the excited atoms of System 1 wobble, they bump into each other, and any one can pass its energy on to any of its neighbors. If this happens, then the first atom turns OFF, and the second turns ON. The newly ON atom is itself now wobbling and jostling its neighbors, and so it may exchange energy with them. The energy, the ON-ness, therefore wanders aimlessly through the system and may arrive at its edge.

At the edge where System 1 touches System 2, the jostling takes place just as it does inside the system itself. An excited atom on the face of System 1 can jostle an atom on the face of System 2, and turn the latter ON. This ON atom jostles its neighbors, and so the energy migrates at random into and through System 2. In this way the thermal motion of the atoms in System 2 is stimulated, but at the expense of System 1. That is, System 2 is *heated* by System 1, and the latter cools (see top figure on next page).

Some time after the initial stage, the energy is spread more uniformly over all the atoms as a result of their jostling each other. The small block still has a higher proportion of its atoms ON than the bigger block, and so it is still hotter. The temperature of System 1 is now 0.72 and that of System 2 is 0.23.

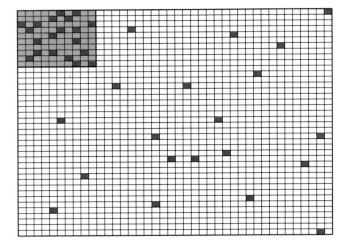

What is the final state of the universe? There is *no* final state for the careful observer, for the ON-ness jostles and migrates forever (there is no rule that brings it to an end). But there is an *apparent* final state for an observer who stands so far back that the behavior of the individual atoms cannot be discerned. There is a final state for the *thermodynamic observer*, not for the atomic individualist. This *apparent* end of change occurs when there is a *uniform distribution of ON-ness*, as in the figure below.

Later, the jostling of the atoms results in a uniform distribution of the energy. There will be small accumulations here and there (there are fluctuations), but on average the proportion of atoms ON in the smaller block is equal to the proportion ON in the larger. The temperatures of the blocks are now the same, at 0.27, and they are at thermal equilibrium.

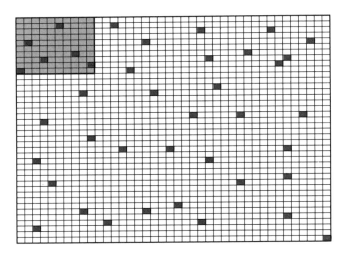

The sequence of illustrations on the two preceding pages shows how the universe attains not so much a final state as a *steady state*. In this state the individual atoms turn ON and OFF as they have always done, but, to the casual observer of averages, the redistribution of energy leaves the universe apparently unchanged. We see that the jostling, random migration of energy disperses it. When it is uniformly dispersed over the available universe, it remains dispersed.

That last remark is not quite true, because the random wandering of ON-ness may lead it to accumulate, by chance, in System 1 and leave Sytem 2 completely OFF. However, even with a universe of 1,600 atoms this change is slight, and in a real Universe, where each system is a block of Avogadro's numbers of atoms, the chance is so remote that it is negligible. *Lack of rules allied with vastness of domain accounts for the virtual irreversibility of the process of dispersal.*

Temperature

Before we wrap this observation into a neat package, let us notice that we are also closing in on the significance of *temperature*. We have just seen that System 1 heats System 2 as a natural consequence of the dispersal of energy, and that the net transfer of energy continues until, on average, the energy is evenly dispersed over all the available atoms. Now note the following important distinction. When the ON-ness is evenly distributed, there is more *energy* in System 2 than in System 1 (because the former contains more atoms, and therefore more are ON when the ON-ness is uniformly distributed), but the *ratio* of the numbers ON and OFF is the same in both.

All this conforms with common sense about hot and cold so long as we interpret the ratio of the numbers of atoms ON and OFF as indicating *temperature*. First, we know that energy flows as heat from high temperatures to low, and we have seen that System 1 (which initially has a higher "temperature" than System 2) heats System 2. Second, the steady state, when there is no net flow of energy between the two systems, corresponds to their having equal "temperatures," not equal total energies. Finally, "temperature" measures the *incoherent* motion, not the coherent motion, of particles; it is intrinsically a thermodynamic (as distinct from a dynamic) property of systems of many particles. It would be absurd to refer to the temperature of a single particle. When we say that a baseball is warm, we are referring to the excitation of its component particles, not to the whole baseball regarded as a single particle.

Temperature reflects the ratio of the numbers of atom ON and OFF: the higher the ratio, the higher the temperature. This interpretation carries over into the actual Universe, where high temperatures correspond to systems in which a high proportion of particles are in excited states. Notice once again the sharp distinction between the temperature of a system and the energy it may possess: a system may possess a large quantity of energy, yet still have a low temperature. For instance, a very large system may have a low proportion of its atoms ON, and therefore be cool; but there are still so many atoms that the sum of their energies is large, and so overall the system possesses a lot of energy. The oceans of the Earth, although they are cool, are immense storehouses of energy. The energy of a system depends on its size; the temperature does not.

A further point is more in the nature of housekeeping. The concept of temperature entered thermodynamics along a classical route; it would be a remarkable piece of luck if the classical definition had turned out to be numerically the same as the ratio of numbers of ON and OFF. The best we can reasonably expect is that increasing temperature in the classical sense corresponds to increasing ratio of ON to OFF in the atomic sense. We cannot assume they will increase in precisely the same way; temperature might increase as the square (or some other increasing function) of the ratio, and not directly as $Number_{ON}/Number_{OFF}$ itself.

We have seen that classical thermodynamics, speaking about the efficiencies of engines, imposes a lower bound to temperature: there is an absolute zero of temperature. In the atomic interpretation, we would expect this to correspond to a system in which no atoms at all are ON (as in the initial state of System 2, depicted in the figure on page 53). Since there cannot be fewer than no atoms ON, the atomic approach to temperature neatly corroborates the classical expectation of a lower bound to temperature. This is just one example of how the atomic interpretation leads to straightforward explanations of classical conclusions.

It turns out that the thermodynamic and atomic versions of temperature coincide in all respects—the temperature goes up if the ratio goes up; the temperature is zero if no atoms are ON; the thermodynamic expressions relating temperature, energy, and entropy all work—if temperature is related to the ratio by

$$Temperature = A/\log (Number_{OFF}/Number_{ON}),$$

where A is a constant that depends on how much energy is needed to turn an atom ON. If for convenience we arbitrarily set A equal to unity, then the temperatures given by this expression are pure numbers (we could choose A to have the units of kelvins, as described in Appendix 2, but that is an unnecessary complication here).

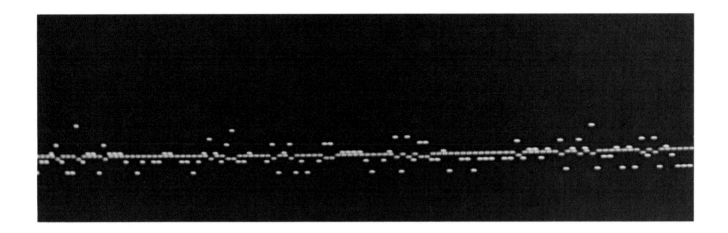

The fluctuations of the temperatures of the two blocks: the yellow points refer to the large system, the green to the small. The calculation has been done on the assumption that 100 atoms of the universe are ON, and the location of each ON-ness is random from one moment to the next. Note that the larger (1,500 atom) system has smaller fluctuations. The two temperatures fluctuate around the same mean in each system. In a system of ordinary size, the fluctuations would be insignificantly small.

By evaluating the preceding formula for appropriate values of numbers ON and OFF, we can ascribe temperatures to the states of the universe illustrated in the figures on pages 53 and 54. The important point to notice is that the temperatures of the two systems converge to a value intermediate between their two initial values. Then, once they have reached equality of temperature, there they remain, except for chance fluctuations, forever. Using a computer, we can map the fluctuations: they are shown in the figure above. Sometimes they are quite large, especially for the smaller system (and so we would notice it jumping between hot and cold around some average temperature); but that is because the systems, and especially the fragment we are calling System 1, are so small. A much larger system would show far smaller fluctuations of temperature at equilibrium; an infinite system would show virtually none.

The Direction of Natural Change

In a very natural way, without imposing superfluous rules, and whittling regulations to the bone, we have arrived at an explanation of the direction of at least one natural change. We have stumbled across one wing of the Second Law. Simply by accepting that jostling atoms pass on their energy at random, we have accounted for one class of phenomena in the world. In fact, this identification of the chaotic dispersal of energy as the purposeless motivation of change is the pivot of the rest of the book. The Second Law is the recognition by external observers of the consequences of this purposeless tendency of energy.

A preliminary working statement of the Second Law in terms of the behavior of individual atoms is therefore that *energy tends to disperse*. (We shall refine the statement as we assemble more information.) This is not a *purposeful* tendency: it is simply a consequence of the way that particles happen to bump into each other and in the course of the collision happen to hand on their energy. It is a tendency reflecting unconstrained freedom, not intention nor compulsion.

At this stage we have seen only the tip of the iceberg of the processes involved in natural change, but as we go on we shall increasingly come to recognize that the simple idea of energy dispersing accounts for all the change that characterizes this extraordinary world. When we grasp that energy disperses, we grasp the spring of Nature. It should also be more apparent now that the Second Law is a commentary about events that are intrinsically simpler than those treated by the First Law. The latter is concerned with establishing the *concept* of energy, something that (it seems to me) remains elusive even after we have analyzed it into its kinetic and potential contributions: after all, what are *they*? Perhaps that elusiveness is appropriate for a concept so close to the Universe's core. The Second Law, on the other hand, regards the energy as an established concept, and talks about its dispersal. Even though we might not comprehend the nature of energy, it is easy to comprehend what is meant by its dispersal.

The interpretation of the Second Law that we have now partially established relates to the Clausius statement, which denies the possibility that heat will travel spontaneously up a temperature gradient. The dispersal interpretation simply says that energy *might* by chance happen to travel in such a way that it ends up where there is already a higher proportion of atoms ON (see below), but the likelihood is so remote that we can dismiss it as impossible. But what of the Kelvin statement of the Second Law: how is dispersal related to the conversion of heat to work?

The migration of energy into a region of the universe that already has a higher proportion of its atoms ON is very unlikely. This is the microscopic basis of the Clausius statement of the Second Law.

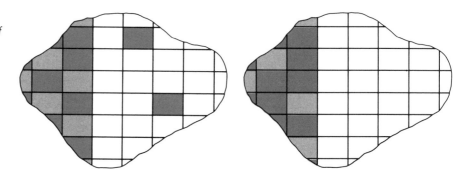

In order to capture the interconversion of heat and work, we have to remember their distinction: work involves coherent motion; heat involves incoherent motion. The atomic basis of the dissymmetry in their freedom to interconvert can be identified by thinking about a familiar example.

Think of a bouncing ball. Everyone knows that a bouncing ball eventually comes to rest. No reliable witness has ever reported, and almost certainly no one has ever observed, the opposite change, in which a resting ball spontaneously starts to bounce, and then bounces higher and higher. This would be contrary to the Kelvin statement, because if we caught the ball at the top of one of its bounces, we could attach it to some pulleys, and lower it to the ground (see below). In doing so we could extract its energy as work. Since that energy came from the warmth of the surface on which it was initially resting (because we are not questioning the validity of the First Law: this is Jack's game again), we would have succeeded in converting heat into work, and the ball is back where it began. That possibility is denied by the Kelvin statement. Therefore, if we can use the idea of the dispersal of energy to account for the absence of reliable reports of balls that spontaneously bounce more vigorously, we shall capture Kelvin as well as Clausius in our net.

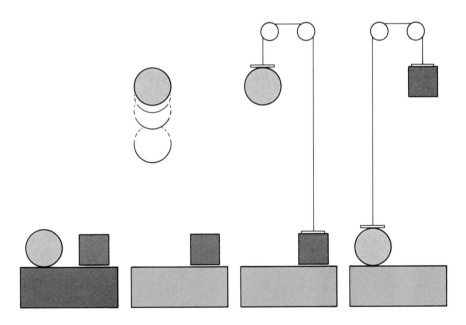

The steps involved in harnessing a bouncing ball for work. On the left, the ball and a weight rest on a warm table. The ball bounces up at the expense of the energy stored in the thermal motion of the particles of the table. At the top of its flight, it is captured and joined to the weight by a cord. As the heavy ball falls, it raises the weight. The overall process is the raising of a weight (that is, work has been done) at the expense of some heat, which contradicts the Kelvin statement.

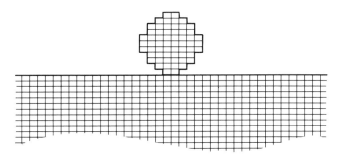

The model of the bouncing ball in the model universe.

A bouncing ball consists of a collection of coherently moving atoms. The bundle of atoms moves coherently upward, slows down, changes direction, and moves coherently downward. If the ball is warm, the atoms also possess energy by virtue of their thermal motion, but we need not trouble about that initially. The motion of the ball can be modeled in the universe as shown above.

In the collision when the ball strikes the table, energy is transferred between the two sets of atoms (and among the atoms of each set). As a result, the atoms of the ball reverse their direction, rise off the table, and climb away. As they do so, their kinetic energy converts to potential; the ball gradually slows, then turns, and drops again.

However, not all the kinetic energy that the ball possessed immediately before the collision remains in it in the form of coherent motion. Some of this energy jostles out while its atoms are in contact with those of the table, and even some that remains becomes randomized in direction. How this happens even in a head-on collision is illustrated in the figure on the left, which shows that what to the ordinary observer seems to be head-on, in fact, on an atomic scale, involves various particles approaching each other over a wide range of angles, and so the motion is transferred in random directions. (A coherent motion as well as a chaotic motion, of the atoms of the table will also be stimulated, because at the point of contact of the ball the atoms are pushed together, and a band of compression travels through the solid. Nevertheless, this band of squashed atoms gets randomized as it moves, and in due course decays into thermal motion. A similar fate awaits the compression wave through the surrounding air, the wave that gives rise to the sound of the ball hitting the table.)

The upshot of this discussion is that each time the ball hits the table, the coherent motion of its atoms is slightly *degraded* into the thermal motion of its atoms and the atoms of the rest of the universe. This is shown by the

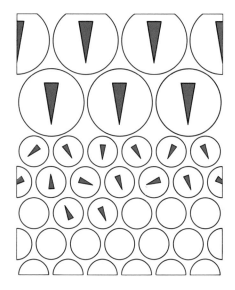

The moment of impact of the ball with the table. At a microscopic level, the impacts are not all head-on, and some of the coherent motion is degraded into thermal motion.

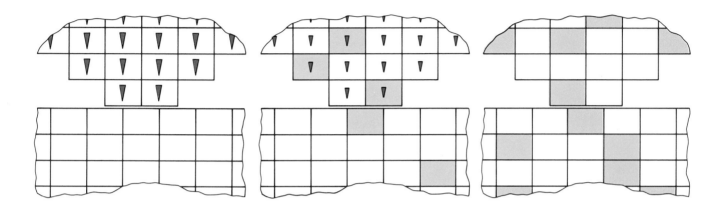

Successive bounces of the model ball. Initially (left) the motion is coherent. After the first impact (middle), some of the coherently stored energy has become incoherent. After several bounces, all coherence has been lost, and the energy is stored as incoherent thermal motion, with ON-ness uniformly distributed over the entire universe of ball and table.

atoms turning ON in the ball and the surface (denoted by the yellow blobs above). As they turn ON the coherent motion gradually turns OFF. Therefore, after each bounce the surface and the ball are a little warmer because the impacts have stimulated the thermal motion at the expense of the coherent. The coherent motion of the atoms of the ball gradually degrades into the incoherent motion of the atoms of the universe. If we wait long enough, all the original coherent motion will have degraded into incoherent motion, and that incoherent motion will be uniformly distributed throughout the universe. The slightly warmer ball will be at rest on the slightly warmer table; moreover, the ball and table will be at the same temperature, for the ON-ness will have dispersed uniformly. The kinetic energy of the ball has been *dissipated* in the thermal motion. Coherence has collapsed into incoherence.

The reverse of this sequence is exceedingly unlikely to occur naturally. We can think of the ball sitting on the warm table. Its atoms and those of the table are wobbling around their mean positions, and there is plenty of energy to send it flying up into the air. However, there are two reasons why the energy is not available.

One problem is that the energy is distributed over all the atoms of the universe. Therefore, in order for the ball to go flying off upward, a good proportion of the dispersed energy must accumulate in the ball. This is not particularly likely to occur, because the ON-ness of the atoms is wandering around at random, and the chance of enough of it being in the ball *simultaneously* is very slight. In a real Universe, with so many atoms, we would probably need to wait a good fraction of eternity before seeing even such an insignificant miracle as a spontaneously bouncing ball, and matter would almost certainly decay first.

But in fact the reasons go deeper, for the ball could be put on a *hot* surface. We have already seen that a 1-kilogram block of iron can rise 1 meter above the surface of the Earth if we transfer 10 joules of energy to it, and such a quantity of energy can readily wander in if the block stands on a slightly hotter surface. But even cool balls placed on hot surfaces do not rise spontaneously into the air. Why? Because the accumulation of energy in the ball, the ON-ness of its atoms, is only a *necessary* condition for it to be able to rise into the air: it is not a *sufficient* condition. In order for the ball to rise, the atoms must be not merely ON, but ON*; that is, the energy must be present as *coherent* motion of the atoms, not merely as incoherent thermal motion. Even if sufficient energy were to wander into the ball from the surroundings, it would be exceedingly unlikely to switch all the atoms ON* and induce coherent motion.

Now we are at the nub of the interpretation of the Kelvin statement of the Second Law. The concept of dispersal must take into account the fact that in thermodynamic systems the coherence of the motion and the location of the particles is an essential and distinctive feature. *We have to interpret the dispersal of energy to include not only its spatial dispersal over the atoms of the universe, but the destruction of coherence too.* Then *energy tends to disperse* captures the foundations of the Second Law.

Natural Processes

The natural tendency of energy to disperse—that is, to spread through space, to spread the particles that are storing it, and to lose the coherence with which the particles are storing it—establishes the direction of natural events. The First Law allows events to run contrary to common experience: under its rule alone, a ball *could* start bouncing at the expense of cooling, a spring *could* spontaneously become compressed, and a block of iron *could* spontaneously become hotter than its surroundings. All these events could occur without contravening the conservation of energy. However, none of them occurs in practice, because although the energy is present it is *unavailable*. Energy does not, except by the remotest chance, spontaneously localize and accumulate in a large excess in a tiny part of the Universe. And even if energy were to accumulate, there is little likelihood that is would do so coherently.

Natural processes are those that accompany the dispersal of energy. In these terms it is easy to understand why a hot object cools to the temperature of its surroundings, why coherent motion gives way to incoherent, and why uniform motion decays by friction to thermal motion. It should be just as easy to accept that, whatever the manifestations of the dissymmetries identified by the Second Law, they are aspects of dispersal.

As energy collapses into chaos, the events of the world move forward. But in Chapter 2 we saw that change is accompanied by an increase of entropy. Entropy must therefore be a measure of chaos. Moreover, we have seen that the natural tendency of events corresponds to the corruption of the quality of energy. Consequently, *quality* must reflect the absence of chaos. High-quality energy must be undispersed energy, energy that is highly localized (as in a lump of coal or a nucleus of an atom); it may also be energy that is stored in the coherent motion of atoms (as in the flow of water).

We are on the brink of uniting these concepts. We have a picture of what it means for the spring of the world to unwind; now we must relate this picture to the entropy. As we do so, we shall acquire Boltzmann's vision of the nature of change.

4 THE ENUMERATION OF CHAOS

$$S = k. \log W$$

LVDWIG
BOLTZMANN
1844–1906

Boltzmann's tombstone in the central cemetery of Vienna. The equation in the inscription reads S = k log W.

Carved on a tombstone in the central cemetery in Vienna is an equation. It is not only one of the most remarkable formulas of science, but also the ladder we need to climb from the qualitative discussion of the dispersal of energy up to the quantitative. The tombstone marks Boltzmann's grave. The formula to the left is our ladder and his epitaph.

Boltzmann's epitaph summarizes most fittingly his work. The letter S denotes the entropy of a system. The letter k denotes a fundamental constant of Nature now known as *Boltzmann's constant* (in what follows we do not need its actual value; so we shall pretend it is equal to unity). The letter W is a measure, in a sense that we shall shortly unfold, of the chaos of a system. Here is our first encounter with a formula that has as many implications for the modern world as Einstein's $E = mc^2$ (the only other equation that people in general seem prepared to know).

Boltzmann's equation is central to our discussion because it relates entropy to chaos. On its left we have the entropy, the function which entered thermodynamics in the train of the Second Law and which is the classical signpost of spontaneous change. On the right we have a quantity that relates to chaos because it measures the extent to which energy is dispersed in the world; the concept of energy dispersal, as we have just seen, is the heart of the microscopic mechanism of change. S stands firmly in the world of classical thermodynamics, the world of distillations of experience; W stands squarely in the world of atoms, the world of underlying mechanism. Boltzmann's tomb is the bridge between the world of appearance and its underworld of atoms.

As Chapter 2 refined the observations discussed in Chapter 1 that gave rise to the perception that energy possesses quality as well as quantity, so this chapter will refine the qualitative discussion of dispersal we met in Chapter 3. Clausius himself saw what we have already seen: he saw the difference between heat and work, understood the intrinsic incoherence of thermal motion, and appreciated what was meant and what was implied by degradation and dispersal of energy. But the world is indebted to Boltzmann for refining that view into an instrument as sharp as a Japanese sword, and showing us how to cut numbers from chaos.

The program of this chapter is to extend and sharpen the blade we have begun to form: we have to enumerate chaos and see *numerically,* rather than merely intuitively, that natural events represent collapse into chaos and that, in a quantitatively precise sense, events are motivated by corruption.

Boltzmann's Demon

How can we quantify chaos? What is the meaning of W? We can arrive simultaneously at both answers by considering a special initial case of the Mark I universe and allowing it to run through its subsequent history. The special initial state is shown below: every atom in System 1 is ON; every atom of System 2 is OFF.

A state of the Mark I universe in which all atoms of System 1 are ON, and all atoms of System 2 are OFF.

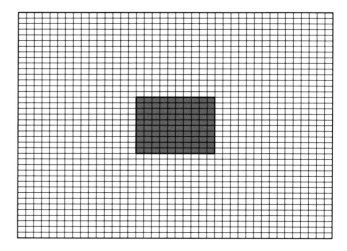

The question we now ask is the following: *how many ways can the inside of a system be arranged without an external observer being aware that rearrangements have occurred?* The answer is what is meant by the quantity W. Notice how this captures what we have earlier called the essential step in going from atoms to systems, an observer's blindness to individuals. Thermodynamics is concerned with only the average behavior of great crowds of atoms, and the precise role being played by each one is irrelevant. If the thermodynamic observer doesn't notice that change is occurring, then the state of the system is regarded as the same: it is only the minutely precise observer who insists on scrutinizing individual atoms who knows that change is actually in progress.

We shall imagine a Demon, a little, insubstantial, neuter, mischievous, and eternally busy thing. I shall call it *Boltzmann's Demon**. Its busyness consists of forever reorganizing. In the universe it simply rearranges ON-ness and OFF-ness. It is the incarnation of the lack of rules that rules the universe. Being infinitely disorganized, all it does is to relocate ONs at random, moving them perpetually but aimlessly.

We, the thermodynamically shortsighted observer, cannot see the Demon. However furiously it reorganizes, so long as it does not change the number of ON atoms in a system (and we note that it can only *move* ON-ness, not create it), then we cannot see that it is active, or even that it is there. Boltzmann's W, then, is the number of different arrangements his Demon can stumble into without us being aware that changes are afoot. If, however, the Demon does manage to move an ON out into System 2, then we shall know that it has happened: the temperature of System 1 will have dropped, and that of System 2 will have risen. We the shortsighted can see our thermometers.

In the special initial state of the universe that we are considering, the Demon cannot do anything without our noticing. All the atoms are ON in the system, and so ON-ness cannot be shifted around within it. Since there is only one arrangement possible in which all the atoms in System 1 are ON, we conclude that $W = 1$. Since the logarithm of unity is zero, Boltzmann's equation gives us the entropy of this state of System 1 as zero. There is *zero entropy* in this highly localized, tidy collection of energy; so it has perfect quality.

In due course the Demon will succeed in moving one ON-ness into the other system (see the figure on the following page). This is the dawn of the Demon's day. Now it can rearrange the ON-ness within System 1 in many different ways, and we the external observer will be none the wiser. It is quite easy to calculate the new value of W: it is equal to the number of different ways of choosing which atom is to be OFF. There are 100 atoms in System 1, and as the Demon moves 99 ON-nesses around, there are 100 places for the location of the one OFF. That is, $W = 100$: this state of the system is one that the Demon can arrange in 100 different ways. Then,

---

* J. C. Maxwell had a Demon too, *Maxwell's Demon*. Its mischief is quite different from Boltzmann's Demon's, and the two should not be confused.

One ON-ness has escaped from System 1 into System 2: the resulting OFF in System 1 can be in 100 different places; the single ON in System 2 can be in 1,500 different places.

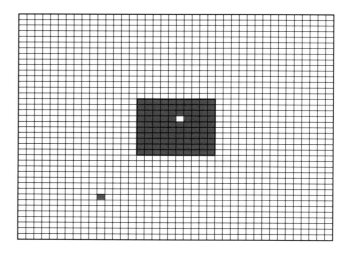

since the natural logarithm* of 100, ln 100, is 4.61, Boltzmann's epitaph gives the entropy of this state as 4.61. The entropy of System 1 is greater than before; the system is more chaotic because we do not know the location of just one OFF.

In due course the Demon will succeed in turning another atom OFF in System 1 and transferring its energy to an atom of System 2. Now there are two gaps in the ON-ness of System 1, and the Demon has more scope for its invisible mischief. The number of ways of arranging the 98 ON-nesses it has at its disposal in System 1 is the same as the number of arrangements of the two OFFs it now must have there. One of these OFFs can turn up at any of the 100 sites; the second can turn up at any of the remaining 99 sites (see the figure on the facing page). Therefore the total number of arrangements of ON-ness that the Demon can succeed in stumbling into is 100 × 99 = 9,900. However, some of these arrangements are identical. For instance, the Demon could first turn OFF atom 23 and then turn OFF atom 32, or it could first turn OFF atom 32 and then atom 23. The end result in each case is the same: atoms 23 and 32 are OFF. Therefore we should divide the previous number by 2, because only half the 9,900 arrangements are different. This means that $W = 4,950$, and that the Demon has 4,950 differ-

---

* The expression log x is now conventionally understood to mean a logarithm to the base 10. We always work here with *natural* logarithms, logarithms to the base e; e is a certain irrational number, 2.78 . . . , that is, a nonrepeating decimal like π. It was arrived at because it enables a great simplification of important types of calculations.

At this stage, two ON-nesses have stumbled out into System 2. There are (100 × 99)/2 different ways of distributing the two resulting OFFs in System 1, and (1,500 × 1,499)/2 different ways of distributing the two ONs in System 2.

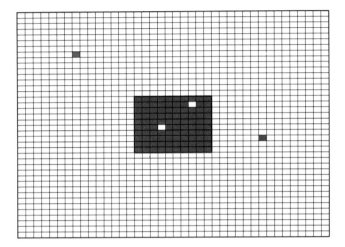

ent ways of reorganizing System 1 without us knowing that anything is going on. Using Boltzmann's tomb, we find that the entropy of System 1 has risen to ln 4,950 = 8.51.

We must not forget that the entropy of System 2 is increasing. Initially it was zero, because no atom was ON, and there is then only one arrangement. Then, when the Demon happened to ship out one ON-ness from System 1 to System 2, one atom turned ON. In System 2 there are 1,500 locations for ON, and so the number of undetectable and indistinguishable ways of arriving at this thermodynamic state of System 2 is 1,500; its entropy therefore is ln 1,500 = 7.31. When there are two ON-nesses to accommodate one can be in 1,500 locations, the other in any of the remaining 1,499. Again, we must not double-count; so the total number of different arrangements is half of 1,500 × 1,499, or 1,124,250. This is the number of different ways in which the thermodynamic state of System 2 can be achieved. The entropy of this state is the logarithm of this number: ln 1,124,250 = 13.93. Notice that the entropy of System 2 is increasing more rapidly than the entropy of System 1: because System 2 is larger than System 1, a single ON-ness in System 2 can be located at more sites than in System 1: the Demon has more scope for rearrangement when it has more atoms to turn ON and OFF.

We could continue to calculate the numbers of arrangements that the Demon can explore, and then take logarithms to arrive at the corresponding entropies. Numbers get very large, but the advantage of taking logarithms is that they cut big numbers down to small: logarithms are very lazy

The entropies of System 1, System 2, and the total universe for different numbers of ONs escaped from System 1 into System 2. The entropy of the universe reaches a maximum (of magnitude 369) when the number in System 2 is between 93 and 94. The temperatures of the two systems are then the same.

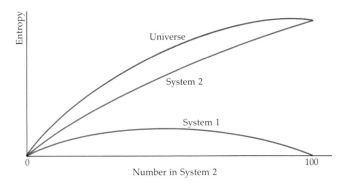

numbers. (For instance, the natural logarithm of 100 is 4.61; the natural logarithm of Avogadro's number is 54.7, even though the number itself is more than 10^{23}.) Therefore, although numbers of arrangements may become astronomical, the corresponding entropies remain terrestrial.

The values of the entropy of each system and of the universe (their sum) are shown above. The entropy of System 1 initially rises, because the Demon has more freedom to locate the ONs as soon as gaps are available; but as soon as half the atoms are turned OFF, the entropy begins to fall, because now the Demon is running short of ONs. If all the atoms were to be extinguished, the Demon would be unable to act; so the entropy would again be zero. The entropy of the other system behaves differently: although it is gaining energy, it will never acquire enough to turn one half of its atoms ON (there are only 100 ONs initially, whereas System 2 has 1,500 atoms). Therefore the entropy of System 2 only rises. The entropy of the universe as a whole therefore goes through a maximum.

From the graph we see that the maximum of the universe's entropy occurs when the proportion of atoms ON to OFF in System 1 is the same as that in System 2, that is, when their temperatures are the same. This is exactly what we expect the entropy to signify. We have seen intuitively that energy will disperse, and we know that this dispersal must correspond to the increase of the universe's entropy. Now we have seen that Boltzmann's epitaph captures both wings of description: "energy tends to disperse" is equivalent to saying that "entropy tends to increase."

Notice too how the illustration at the top of the facing page lets us account for the natural direction of energy flow in a temperature gradient. Suppose we have an initial arrangement of the universe in which only one

An initial state of the Mark I universe in which 99 atoms are ON in System 2 and only 1 is ON in System 1.

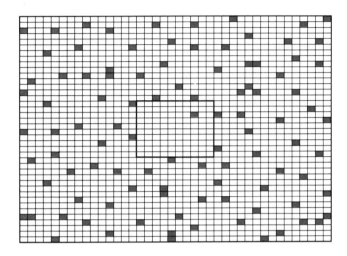

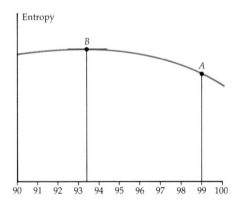

The entropy of the Mark I universe shown in the figure at the top of the page, with increasing numbers of atoms OFF in System 2 (and ON in System 1). The initial state (in the figure at the top of the page) corresponds to the point marked A: the maximum entropy is reached at B. At B the two systems are in thermal equilibrium, and their temperatures are the same (apart from fluctuations).

atom of System 1 is ON, and 99 atoms of System 2 are ON. Then we know from our earlier remarks that the temperature of System 1 is lower than that of System 2 (the temperatures are respectively 0.22 and 0.38 if we use the formula given on page 56). The entropy of the universe is therefore at the point marked *A* in the figure on the left. *Intuitively* we know what will happen: the energy of System 2 will jostle into System 1 until it is uniformly distributed over the entire available universe (see the figure on the next page). This corresponds to each of the 1,600 atoms having an equal likelihood of being ON: since there are 100 ONs overall, at equilibrium we can predict that the chance of any one being ON is 100/1,600, or 0.0625, whether the atom belongs to System 1 or to System 2. Since there are 100 atoms in System 1, the number of its atoms ON at equilibrium is 100 × 0.0625 = 6.25. However, that number must be an integer because atoms are only fully ON or OFF; therefore the number must be fluctuating around 6 and 7; for simplicity we take it to be 6 (or occasionally 7). The other 94 (or 93) ON atoms are therefore all in System 2.

When 6 (or 7) atoms are ON in System 1, the temperature is 0.36 (0.39); when 94 atoms are ON in System 2, its temperature is 0.37 (it is also 0.37 when 93 are on, because in the bigger system the temperature is less sensitive to numbers). These temperatures are virtually the same (the difference arises from the fact that we have rounded 6.25 to 6 or 7). Not only are the temperatures the same, but they correspond (as we can see from the figure to the left) to the maximum value of the entropy of the universe, point *B*, exactly as our earlier discussion requires: the cooling to thermal equilibrium corresponds to an increase toward maximum entropy.

One distribution of ONs corresponding to thermal equilibrium between Systems 1 and 2, and to point B *on the entropy curve.*

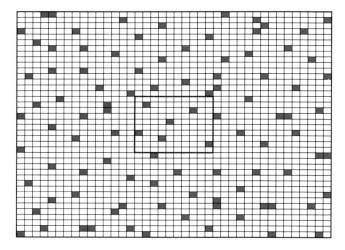

The Demon's Cage

Each successful quantitative step of science brings in its wake new qualitative insight. The progress of science can often be traced to a symbiosis of insight and mathematics: each one eases the other along, and as progress is made so comprehension flourishes. The same is true of the step we have now taken: the step, from the intuitive notion of chaos to its precise formulation in terms of the number of arrangements open to a system but invisible to an external observer.

The new insight obtained from Boltzmann's tomb concerns the nature of equilibrium. In the model we have been considering, the maximum entropy of the universe occurs when the two systems are at thermal equilibrium. Then there is no net flow of energy from one to the other, and there the two systems will remain forever, except for chance fluctuations that happen, very occasionally, to ripple the evenness of the distribution. At thermal equilibrium the systems appear to be at rest, and net change is quenched. But in fact the Demon is as active as ever. Boltzmann's Demon never dies; it scurries furiously and randomly from atom to atom, extinguishing here and igniting there. Thermal equilibrium is an example of *dynamic equilibrium*, where the underlying motion continues unabated and the externally perceived quiet is an illusion. Almost all the final resting conditions of the processes that we shall consider are dynamic equilibria of this kind, and we shall see many examples of atomic life continuing after the bulk seems dead.

But there is an even more important point. Dynamic equilibrium represents the Demon caught in the cage of its own spinning. Thermal equilibrium, as we have seen, corresponds to the condition of maximum universal entropy. It therefore also corresponds to the thermodynamic (average) state that can be achieved in the maximum number of ways. If we think of the universe as being able to exist with many arrangements of ONs scattered over either system, then different scatterings may correspond to different thermodynamic states; but in general many different scatterings of ONs will correspond to each state. We can then ascribe a *probability* to each thermodynamic state in terms of the number of ways in which, at a microscopic level, it can be achieved. Then *the more ways in which a state can be achieved, the higher its probability*, in the sense that a chance scattering of ONs is more likely to land in an arrangement corresponding to a given thermodynamic state if that state can be achieved in many ways. In this sense the *uniform* distribution (which is also the one that can be achieved in most ways) is the *most probable* state of the universe. In other words, *thermal equilibrium corresponds to the most probable state of the universe.*

This conclusion can be expressed in a slightly different way. We allow the Demon perfect freedom to shift and change; therefore, in due course, it runs through all the possible arrangements of 100 ON-nesses (and may enter many arrangements many times). We may have to wait a trillion years, but the time will come when every configuration of the universe will have been achieved. However, *almost all* the arrangements correspond to a uniform distribution of ON-ness; perhaps for a millisecond in those trillion years the universe will be found with all the atoms of System 1 turned ON, but for most of the time the energy will be almost uniform. This is because there are so many arrangements that correspond to uniformity (but which are imperceptibly different to the onlooker) that the Demon spends most of its time generating them, and for only a miniscule fraction of its time does it happen to achieve others.

Of course, with a universe of only 1,600 atoms and with System 1 being as small as 100 atoms, the chance that significant abnormalities will be stumbled into by the Demon is quite large. Nevertheless, substantial fluctuations occur only infrequently, and most of the Demon's labors are imperceptible. As an example of this kind of behavior, the figures on the next two pages show several frames in succession: all of them correspond to having six or so atoms ON in System 1. Even though the atoms that are ON are different in each frame, we of the blunted thermodynamic eye cannot perceive that. We regard the system as being in a steady state: the thermometer remains steady while the Demon deploys.

This feature of change is exceedingly important. There are many states of the universe, and the random wandering of the energy permits them all, in principle, to be achieved. A fragment of the universe might begin in a

Some more of the myriads of arrangements of ONs at thermal equilibrium. Most of the time there are 6 or 7 atoms ON in System 1.

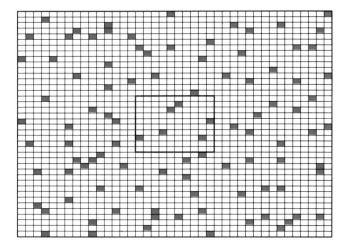

highly improbable state (for example, in the arrangement shown in the top figure on page 71, in which System 1 is relatively cold). After that we shall see the universe drifting through ever more-probable states. That is the natural direction of spontaneous change. When the universe arrives at a more probable state (that is, one that can be achieved in more ways), it almost certainly does not return to a less probable one, because the likelihood of random jostling taking it there by chance is too remote. The final condition of equilibrium of the universe is then its *most* probable state. The Demon has spun its own cage: the very chaos with which it acts ensures that it is trapped in the future and cannot return to the past. It *could* return to the past by unraveling chaos if it acted purposefully; but it acts at random, and chaos cannot undo chaos except by chance.

Such are the properties of the model universe. The properties of our actual Universe mirror them precisely, but its energy can be dispersed in so many ways that extraordinary structures may emerge and appear stable as the Universe sinks virtually irreversibly toward equilibrium. However, we have now discovered the essentially *statistical* way in which systems evolve. We see that the irreversibility of natural change results not from certainty, but from probability: perceived events correspond to the evolution of the Universe through successive states of increasing (and, once attained, overwhelming) probability. In principle, therefore, the Universe has loopholes for miracles.

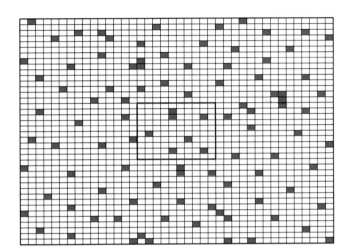 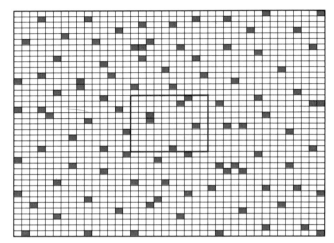

It would be regarded as a minor miracle, for instance, if a lump of metal were suddenly spontaneously to glow red hot, let alone if water were spontaneously to turn into wine. But the Demon might succeed in bringing about at least the lesser miracle, and could do so by chance. It is conceivable, because the probability is not absolutely zero, that the aimless actions of the Demon could accumulate a great deal of energy in a tiny region of the Universe. But the *probability* of that happening is negligible, and the probability that the fundamental particles of water might stumble into an arrangement that we would recognize as wine is even more remotely infinitesimal. The loophole exists, but it is almost infinitely small, and the greater probability is that the reports of miracles are exaggerations, falsely reported rumors, hallucinations, deceptions, misunderstandings, or simply tricks. To paraphrase David Hume: it is always more probable that the reporter is a deceiver than that the miracle in fact occurred.

Chaos, Coherence, and Corruption

The relation of the entropy to W as expressed by the Boltzmann equation sharpens the meaning of chaos. We shall do two things with it. First, we shall express more precisely what happens in the course of natural change. Then we shall use it to encompass the disorder in the way that matter is arranged.

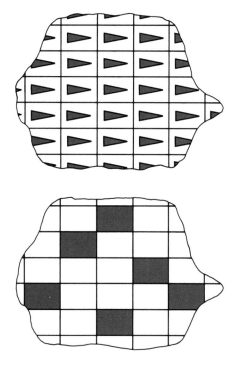

The natural direction of change is from coherently stored energy (upper illustration) *to incoherently stored energy* (lower illustration).

Consider what is involved in the chaotic disruption of coherence, as when the ordered motion of a body (see figure on left) gives way to thermal motion (as we discussed for the bouncing ball). The entropy of the initial state of the body is zero, because all the atoms are moving coherently. In terms of the activities of the Demon, there is no way in which it can rearrange the ON*-ness, for any change would alter the state of motion of the body, which we would detect. Hence Boltzmann's W is equal to unity, and his equation gives an entropy of zero. The body might be warm, in which case it would possess an entropy, but that would merely add a thermal contribution to the total; for simplicity we shall suppose the temperature to be zero, and therefore that there is no entropy from this source. The table that the body is about to strike is also perfectly cold, or so for simplicity we may suppose.

When the perfectly cold body strikes the perfectly cold table, energy is dissipated into the thermal motion of the atoms of both. The entropy of the table and the body therefore both rise, because now the Demon has ONs to deploy. Overall, therefore, there has been an increase of entropy.

The universe is shifting toward a state of higher probability. Initially there is only one arrangement for the ON-ness of the atoms (and indeed they have to be not merely ON but ON* in a definite direction). This coherent motion of correlated excitation would be a very improbable outcome if the Demon were simply handed a bag of ONs and were left to deploy them. On the other hand, each successive bounce leaves the universe in a more probable arrangement, one that the Demon is more likely to achieve. In the end, when the energy is uniformly and incoherently distributed, the universe is in its most probable state, the state in which the Demon can spin arrangements almost forever without detection.

Now we take the last step toward the complete identification of chaos. Suppose that the particles of the universe are free to move, and that they, as well as their energy, can move from place to place, as they could if the universe were a gas. Suppose we prepare an initial state by injecting a puff of gas into one corner of the universe (upper figure on right). We know intuitively what will happen: the cloud of particles will spontaneously spread and in due course fill the container.

That behavior is easy to understand in terms of the onset of chaos. A gas is a cloud of randomly moving particles (the name "gas" is, in fact, derived from the same root as "chaos"). The particles are dashing in all directions, colliding, and bouncing off whatever they strike. The motion and the collisions quickly disperse the cloud, and before long it is uniformly distributed over the available space (lower figure on right). There is now only an extremely remote chance that the particles will ever again simultaneously and spontaneously accumulate back in their original cor-

An initial stage of the Mark I universe is prepared by squirting in a puff of gas (yellow atoms) into one corner.

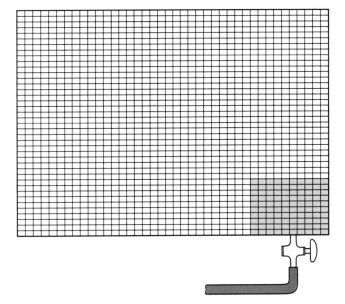

The equilibrium state of the universe consists of particles of gas dispersed uniformly (on average) over the available space. This shows just one such arrangement; there are myriads more. This state of the universe can be achieved in many more ways than the initial state; so the universe is in a more probable state.

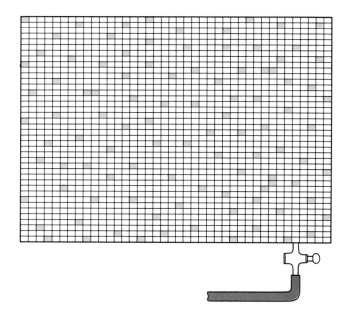

ner. Of course, we could drive them back into the corner with a piston, but that would involve doing work, and the accumulation would not have been spontaneous.

Clearly, the idea that energy tends to disperse accounts for the change we have just described, for now the ON-ness of the atoms has been *physically* dispersed as the atoms themselves spread. Each atom carries kinetic energy, and the spreading of the atoms spreads the energy. But in what sense has the *entropy* increased? We can get the answer from the Boltzmann expression by thinking in terms of the value of W and the activity of the imperceptible Demon.

An initial state of the universe in which all 800 particles of gas lie in the left half of the container.

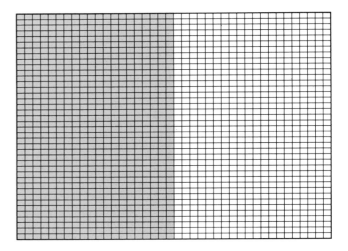

Suppose the initial cloud occupies one-half of the entire universe, as in the figure above. We know, from experience, that in a final equilibrium state the gas will be spread throughout the universe (as in the figure on the facing page), and therefore occupy twice the original volume. In the initial state the Demon's domain is only on the left, and then we know that some particle A must be there. In the final state the Demon can deploy the atoms (which, for convenience, we shall regard as all being equally ON) in either half of the universe. Atom A now may be either on the left or on the right. So long as there are compensating shifts of other particles, as the Demon (now disguised as the chance collisions) moves the atoms from place to place, the external observer is unaware that the inner structure of the gas is a tumultuous storm.

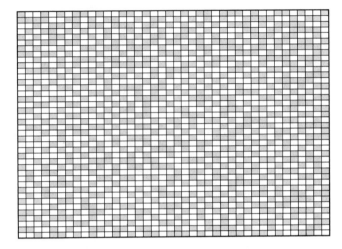

The equilibrium state of the gas, where the particles are distributed uniformly. Now any particle is equally likely to be found in the right half or the left half of the container. The entropy of this state is greater than the initial state by 1600 ln 2.

For each particle, the number of locations the Demon can move it to is increased by a factor of 2 when the gas is allowed to spread throughout the entire universe. Consider two particles: when the second is also allowed to explore the entire universe, it too has twice as many possible locations as it had at first. Therefore, in a sample of two atoms the number of arrangements corresponding to the same energy increases when the cloud expands by a factor of $2 \times 2 = 2^2$. For three the increase is a factor of $2 \times 2 \times 2 = 2^3$, and so it goes on. For a sample of 100 particles the value of W increases by a factor of 2^{100}. Therefore the Boltzmann equation tells us that the entropy increases from its original $\ln W$ to $\ln (2^{100} \times W)$. The increase is therefore the difference of these two quantities,* or $\ln 2^{100}$. This increase is equal to $100 \ln 2$, or 69.3. Hence here also, as we should expect, we have an increase in the entropy of the universe.

The Boltzmann equation therefore captures another aspect of dispersal: *the dispersal of the entities that are carrying the energy.* In fact his tomb is universal. However energy is dispersed, by spreading from one platform to another, or by the platforms themselves spreading and mingling with other platforms, or by a simple loss of coherence within a sample, it corresponds to the increase of entropy. That is the power of the Boltzmann equation: it enumerates corruption in all its forms.

---

* We are using the property of logarithms which tells us that $\log ax - \log x = \log a + \log x - \log x = \log a$. In the next step we use the relation that $\log x^a = a \log x$. The rules are true for logarithms to any base.

5 THE POTENCY OF CHAOS

This is the turning point of the fortunes of chaos. Our hero, apparently committed to a life of dissipation, degeneration, and general corruption, is about to make good.

On the one hand, we have the world of phenomena: the immediate world of appearance and process. This is symbolized by the steam engine. On the other, we have the world of underlying mechanism. This is symbolized by the atom.

Reflection on experiences with the steam engine identified a dissymmetry in the workings of Nature, which, we found, could be encapsulated in the remark that the entropy of the universe always increases in any natural change. Entropy, we saw, was related to the value of *(Heat supplied)/Temperature*. We also saw an economic consequence of the dissymmetry: there is an intrinsic inefficiency, a tax to pay, when heat is converted into work. This inefficiency is governed by the temperatures involved in the operation of an engine.

Reflection on the microscopic world of atoms showed that we should expect natural processes to be those in which there is a dispersal of energy. We have refined the meaning of "dispersal," and have seen that it signifies the spreading of energy, either by the motion of what carries it or by its transfer from one carrier to another. We have also seen that dispersal signifies a loss of coherence in the manner in which energy is stored. We have claimed that all the processes of the world are aspects of this general dispersal, and that spontaneous processes are the manifestation of the purposeless, underlying spreading that chance brings about and that lack of regulation allows.

The bridge between the two worlds is the epitaph on Boltzmann's tombstone (see page 65). It relates the entropy, as encountered in the world of experience, to a measure of dispersal, which we can interpret in terms of events in the microscopic world. The Universe, we have seen, is ineluctably drifting through states of ever-increasing probability. Once any new state has been attained (by any natural action), the Universe is locked out of the past, for any turning back is too improbable to be significant.

That is the general background to the events that surround us and take place within us. But Nature has an extraordinary way of slipping into chaos, and sometimes (often, in fact) does so unevenly. The world does not degenerate monotonously. Here and there a constructive act may effloresce, as when a building or an opinion is formed. The descent into universal chaos is not uniform, but more like the choppy surface of rapids. In a *local* arena there may be an abatement of chaos, but it is an abatement driven by the generation of even more chaos elsewhere.

We now need to unravel the network of connections that Nature drives as it sinks into chaos. We shall begin by returning to the Carnot cycle, newly equipped with our insight into the purposeless behavior of atoms and energy, and see how the corruption of the quality of the energy in the world may bring about local abatements of chaos. Then we shall explore the structural potency of chaos.

Carnot under the Microscope

We can build a simple model of the Carnot engine using the Mark II version of the universe (below). The Mark II universe, remember, is like the Mark I, but the surroundings of the system of interest are infinite: the hot source is an inexhaustible supply of energy, and the cold sink is an insatiable absorber of energy. The indicator diagram for the Carnot cycle we described in Chapter 1 is reproduced above, right. What we now have to

The Mark II universe model of the Carnot engine. A constant amount of gas is contained in the cylinder; the hot source and the cold sink may be brought into contact with the gas, or it may be insulated from them. Compare the figure on page 15.

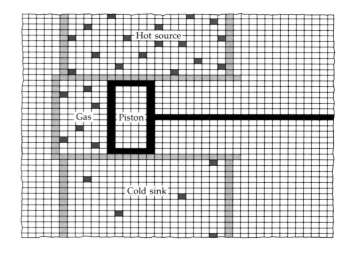

The indicator diagram for the Carnot cycle. (This is the same as in the figure at the bottom of page 18.)

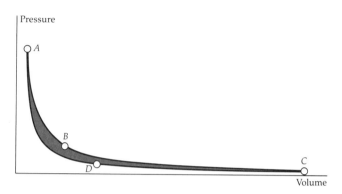

establish is how the random dispersal of energy succeeds in producing coherent motion: we have to establish at a *microscopic level* how heat is converted into work.

At *A*, at the start of the cycle, the working gas is at the temperature of the hot source. That is, the ratio of the number of ON atoms and OFF atoms is the same in both. (We are taking the simplistic view that the energy needed to turn an atom ON is the same in the surroundings as in the working substance.) From now on we shall say simply, "The ON:OFF ratio is the same." The atoms of the gas are free to move and collide with anything that happens to lie in their path.

All the walls except one are rigidly fixed in place. The exception is the piston. The crucial feature of the engine is that it possesses at least one wall that can move in response to the impacts it receives. Here is an essential asymmetry of the engine: it possesses a *directional* response to the impacts it receives. The face of the piston is, in effect, a screen: it picks out and responds to the motion of particles that happen to be traveling perpendicular to it; and it rejects (by not responding to) components of motion that happen to be parallel to it. Engines, in effect, select certain motions of the particles within them. The directionality of the movement of an actual piston in an engine is a consequence of this asymmetry. Our exploitation of heat to achieve work is based on the discovery that the randomness of thermal motion can be screened and sorted by asymmetry of response.

The random thermal motion of the particles of gas is transformed into coherent motion of the particles that constitute the piston (and then of whatever the piston is rigidly attached to). As a result, some of the particles are switched OFF, because they have jostled away their motion (see the top figure on the next page). However, since the gas remains in contact with the hot source as the piston moves back, and since energy continues to

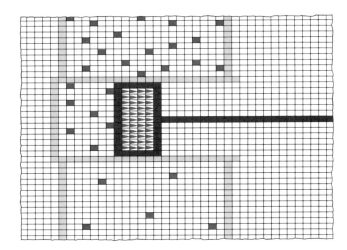

The microscopic events taking place in the iso-thermal power stroke from A to B. The thermal motion of the ON atoms is given up as the co-herent motion of the particles of the piston, but the energy is restored because of the thermal contact between the gas and hot reservoir. En-ergy jostles in, and maintains the temperature (the ON:OFF ratio) of the gas atoms.

jostle in its normal purposeless way, the ratio of numbers ON and OFF actually remains the same. The *A* to *B* leg of the cycle is therefore the consequence of purposeless wandering, with the asymmetry of the environment extracting coherence from random motion.

At *B* the thermal contact with the environment is broken. The piston continues to respond to impacts, and goes on being driven out, but no more energy can jostle in from the hot source (below). In this adiabatic step, therefore, atoms are turned OFF as they give up their energy to the piston. The ON:OFF ratio falls, and with it falls the temperature.

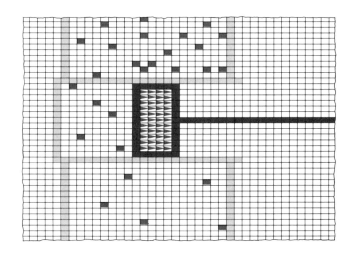

The events during the adiabatic power stroke from B to C. Although the thermal motion con-tinues to be converted into the coherent motion of the particles of the piston, the thermal insula-tion of the gas means that the ON-ness is not restored. The number of atoms ON decreases; so the temperature falls. (In the engine, recall, the number of atoms in the cylinder remains con-stant: this feature is misrepresented by the model.)

The crank has now turned so far that the piston begins to reenter and compress the gas. On a microscopic scale, this means that the coherent motion of the particles of the piston is stimulating the motion of the gas particles: although the stimulated motion is briefly coherent, it rapidly degrades to thermal motion. In other words, the piston turns atoms ON. Since the gas is now in thermal contact with the cold sink, the excess energy jostles away, and the proportion ON (and hence the temperature) remains constant.

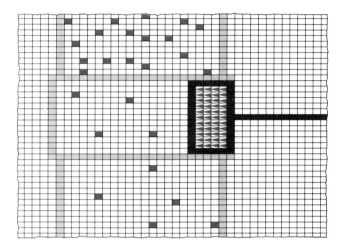

At C the turning of the crank reverses the direction of motion of the piston, and thermal contact is established with the cold sink (above). Now the coherent motion of the incoming piston stimulates the particles to move more rapidly as they collide with it (just as ping-pong balls go faster after being hit by the paddle: compression is just a vast, simultaneous game of ping-pong). Thus *work* is being done on the gas, because energy is being transferred to it by the coherent motion of the particles of the piston. This coherent motion is picked up by the particles of gas. However, the particles collide among themselves so rapidly that in fractions of a second the motion has been rendered incoherent. Although work is being done, the coherence of the motion is dissipated so quickly that it results in incoherence. The gas, however, although it is increasingly turned ON, does not get hotter: the jostling of the atoms among themselves and with the walls ensures that the gas remains at the same temperature as the cold sink with which it is now in contact.

At D the thermal contact with the sink is broken, and the compression becomes adiabatic. The particles of the piston continue to stimulate the motion of the particles of the gas, and more and more of these turn ON (see figure on the next page). Now they cannot jostle their energy to the surroundings; so the work done by the incoming piston raises the temperature of the gas. This brings us to A, and the cycle is complete.

In the course of completing the cycle, more disorder than order has been created. The coherent raising of the weight to which the piston is attached is a process perfectly free of entropy production (so long as it is quasistatic). We draw energy from the hot source. That reduces its disorder, for with fewer atoms ON, the Demon has less scope for rearrange-

In the final, adiabatic step from D *to* A, *the incoming piston continues to stimulate atoms to turn ON. The energy cannot escape because of the thermal insulation; so the number ON (and with it the temperature) rises.*

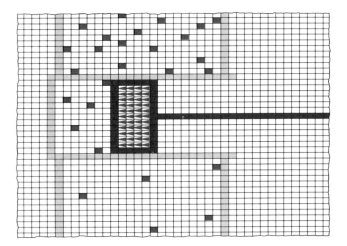

ment. Less energy is dumped into the cold sink, but so long as it is cold enough (so that the ON:OFF ratio is low), the Demon lurking there will gain more opportunities to deploy ONs than the Demon in the hot source loses. That is, even a small supply of energy to a cold sink can generate a lot of chaos (a sneeze in a library has more impact than a sneeze in a crowded street). Therefore, provided *some* atoms are turned ON in the cold sink (and the appropriate number depends on the proportion already ON, that is, on the temperature there), we may be able to produce more disorder in the world than we had originally, even though we have eliminated some disorder from the hot source by withdrawing some energy as heat. Consequently, the running of the engine, and its production of work from heat, is a spontaneous, natural process, and the engine will run forward.

This can be expressed differently. The state of the universe at the end of the cycle is more probable than its initial state (in the sense of Chapter 4 since it can be achieved in more ways). Hence the universe enters the new state spontaneously, and then remains there. Entering this more probable state has resulted in a weight being lifted. The raised weight represents the local abatement of chaos, but it has been raised because chaos has been produced elsewhere.

The cycle may be complete, but the world is no longer the same. Energy has wandered out of the hot reservoir and into the cold, but some of it has raised a weight. The shaft of the engine may have raised bricks, blocks, and girders, and from them may have emerged great cathedrals to either gods or mammon. Yet notice how they have been built. They have been built by *destruction*.

Engines may be coupled to bricks, blocks, and girders and result in the construction of buildings: these constructions, though, are the result of destruction.

Stirling's Engine

The Carnot engine is an abstract design. One reason why it cannot be used to build a practical machine is apparent from the illustration on page 83: the area bounded by the cycle is very small. Although the cycle is efficient (if gone through quasistatically), each rotation of the crank delivers very little work. In the remainder of this chapter, we shall examine some of the cycles that are used commercially: we shall see that each is driven by the generation of chaos, even though ostensibly each is driven by the consumption of fuel. Our trucks, automobiles, and jet airliners are all impelled by corruption.

Robert Stirling was a minister of the church and active during the opening years of the nineteenth century, when people were being killed or maimed by explosions that resulted from the use of increasingly higher steam pressure in engines. The ambitions of the engineers outstripped the capabilities of the metallurgists, as they sought to confine high pressure within the inadequate steels of the time. As befitted his calling, he grieved over such personal tragedy, and was to devise an engine that would work at lower, less-dangerous pressures. The *Stirling engine* remained largely forgotten (like his sermons); but recently it has come of age, for it can be pollution-free, self-contained, and quiet, and has been found especially suitable for refrigeration (when run in reverse).

The principle of the Stirling engine is illustrated below. It consists of two cylinders, each fitted with a piston, and a special device called a *regenerator* (Stirling, a Scot, called it an *economiser*) in the pipe that joins them. The two pistons are connected to a shaft, but in a way so subtle that it long

An idealized Stirling engine. The cylinder on the left is kept hot by a flow of hot fluid (or electrically, or by focused sunshine); the cylinder on the right is kept cold (by a flow of water, for instance). Between them lies the regenerator. The two pistons are connected by a complicated set of gears and cranks that couple their motions in a clever way.

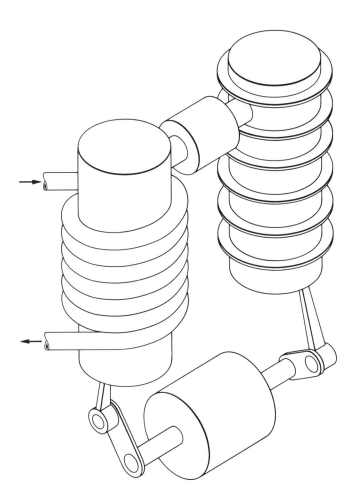

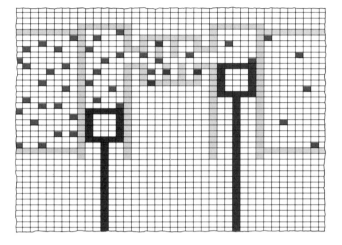

The Mark II model of the Stirling engine. The hot cylinder is in contact with the hot reservoir on the left; the cold is constantly in contact with the cold reservoir on the right. The regenerator lies between them.

defeated the practical implementation of his design. The aim of the connection is to coordinate a complicated sequence of motions of the two pistons, as we shall shortly describe. One of the cylinders is kept hot by a burning fuel or an electric heater. The other is kept cool by cooling vanes or the flow of water. In the model of the engine in the universe (above), one cylinder is permanently in contact with the inexhaustible hot source, and the other is in touch permanently with the insatiable cold.

The regenerator is the special feature of the engine. It consists of a collection of vanes of metal or a pad of wire wool. It has two features. First, it must not be too good a thermal conductor, because it stands between the hot and cold regions of the engine, and the temperature difference must be maintained. Second, it must act as a temporary reservoir, able to absorb heat as hot gas flows through it, and able to give up that heat as cold gas flows through it later. This is its regenerative function: to reheat the cold gas, and to recool the hot.

Initially the engine has its pistons arranged as in the figure at the top of page 90. The piston in the cold cylinder (henceforth piston$_{COLD}$) is fully inserted, and the piston in the hot cylinder (piston$_{HOT}$) is half-way out. Piston$_{HOT}$ moves out while piston$_{COLD}$ stays still. This is the power stroke: the crank is turned, and energy floods in as heat from the hot source, exactly as we have already seen in the Carnot engine. This can be represented in the model universe as depicted in the figure. The step takes us to B. Since the volume of the gas has increased, but its temperature has remained the same, its pressure declines. This is shown in the indicator diagram at the bottom of page 90.

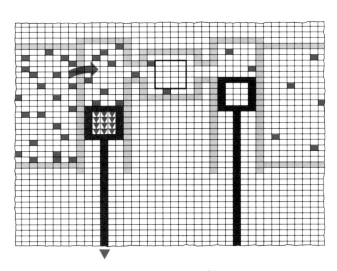

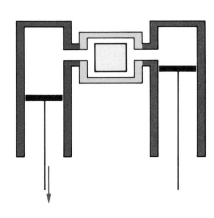

The power stroke of the Stirling engine. In this stage (which corresponds to going from A to B in the indicator diagram), the piston on the left moves out isothermally, and the piston on the right remains still. Energy is drawn in from the hot source, and thermal motion converts to coherent motion of the particles of the hot piston.

At *B* the linkage between the pistons is such that, as piston$_{\text{HOT}}$ moves in, piston$_{\text{COLD}}$ moves out (figure at upper right). This preserves the total volume of the gas as it is shipped from one cylinder to the other. But it is *hot* gas; so as it flows from one cylinder to the other, it heats the regenerator; the gas's ON atoms jostle the atoms of the temporary reservoir. This cooling of the gas at constant volume decreases its pressure, which brings us to *C* in the indicator diagram below.

At *C* Stirling's clever linkage between the motions of the piston keeps piston$_{\text{HOT}}$ stationary as piston$_{\text{COLD}}$ moves in (figure at lower right). This compresses the gas; but the gas's temperature does not increase, because the piston is connected to the cold sink. Energy jostles out, and the pressure of the gas rises isothermally. This takes us to *D*. Notice that we have shipped heat from a hot source to a cold sink.

The indicator diagram for the Stirling cycle. The hot isothermal stage is in red, the cold in blue.

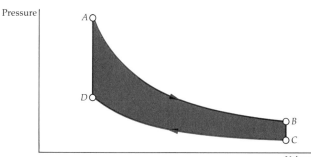

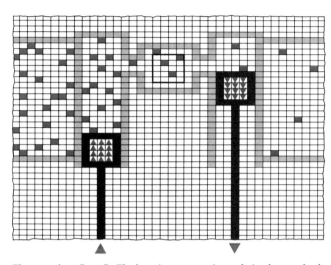 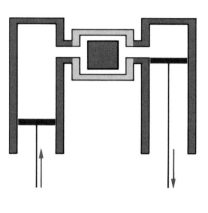

The stage from B to C. The hot piston moves in, and simultaneously the cold piston moves out. This maintains constant volume of the working gas. As the hot gas flows through the regenerator, energy jostles out to turn its particles ON. The gas itself cools.

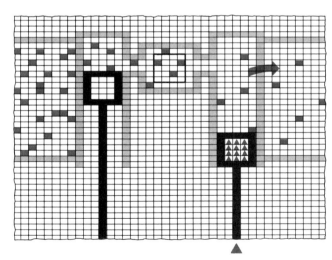 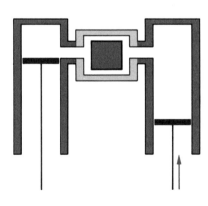

The stage from C to D. Now the hot piston is stationary, and as the cold moves in, the ON-ness it stimulates jostles out into the cold sink. This is an isothermal compression step, and work is done on the cold gas.

The fourth leg completes the cycle. In order to bring the cycle from D to A, piston$_{HOT}$ moves out, and piston$_{COLD}$ moves in (see below). This maintains a constant volume of the gas (hence the line on the indicator diagram is vertical), and ships it from its cold cylinder into the hot one. As the gas passes through the regenerator, it is heated by the energy previously stored there, and thus simultaneously cools the regenerator back to its initial condition. Now we are back to A: the regenerator is once more ready to absorb heat, and the cycle can begin again.

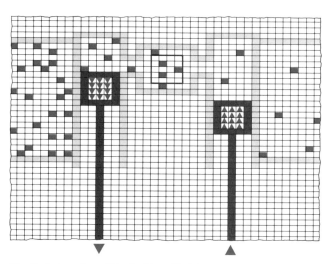

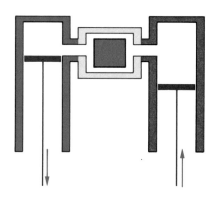

The final stage, from D *to* A. *The cold piston moves in, and simultaneously the hot moves out. Gas is shipped from the cold to the hot cylinders, and acquires ON-ness as it flows through the hot regenerator (which is thus cooled).*

The Stirling and Carnot engines are similar in that each one works by drawing high-quality energy from a hot supply and dumping it into a cold sink: work is achieved at the expense of the corruption of energy. Furthermore, the thermodynamic efficiencies of the engines are the same: if each is working perfectly, and the cycle is gone through quasistatically, the quantity of energy that has to be discarded in order *just* to avoid creating more order in the Universe is exactly the same in each engine. Therefore the efficiency of the Stirling engine is given by the expression derived for the Carnot cycle itself, namely $1 - (Temperature_{COLD}/Temperature_{HOT})$.

There is, however, a difference between the two engines: a larger area is enclosed by the cycle in the indicator diagram for the Stirling engine than for the Carnot engine. The larger area means that each cycle in the Stirling

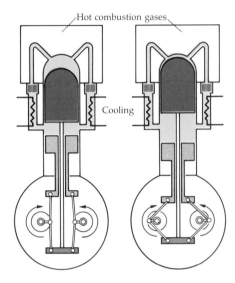

A schematic diagram of a working Stirling engine. The hot (red) and cold (blue) pistons are in line, and connected by the linkages in the crankcase. The regenerator lies between the region for the hot combustion gases and the cooling coils. On the left, the pistons are making their closest approach, and most of the gas is in the hot region. On the right, the pistons are furthest apart, and most of the gas is in the cold region.

engine delivers more work (it has also to absorb more heat; so the efficiency criterion is not contravened). Therefore the Stirling cycle is more suitable for practical applications than the Carnot cycle, because each turn of the crank is more productive.

Productive it may be, but cumbersome it most certainly was, and the first Stirling engines were pretty useless affairs: the linkages between the pistons impaired the efficiency by friction, and the regenerator was far from ideal in its operation. Nevertheless, an engine that can run quietly on any fuel (including sunshine) has obvious advantages. Modern engineering has made the Stirling engine practicable: Stirling engines are now available that can generate 5,000 horsepower. Furthermore, because the Stirling engine works by *external* combustion, the fuel can be burnt completely, and there are fewer polluting emissions.

A typical design might be something like that shown to the left, and the corresponding actual indicator diagram is shown below: it differs from the ideal cycle we have been considering, but its parentage is clear. The entire engine, including the crankcase, must be sealed. One problem with this engine is that what seems to be sealed to human perception is not necessarily sealed to atoms. Hydrogen could be burned in a Stirling engine; but under the high pressures often used, it diffuses through the "solid" metal walls, and must be replaced continuously. It therefore cannot be used in real applications, even though it is an excellent working fluid, with such low viscosity that it undergoes little frictional loss (the Second Law again) as it is shipped back and forth between the cylinders. However, helium's viscosity is comparable to hydrogen's, and is the working fluid used in space-flight applications. In space, the hot source is focused solar radiation, and the cold sink is a radiator radiating out into space on the shady side of the craft. The work generated by solar energy in this way is used to drive a generator.

The indicator diagram of a practical Stirling engine compared with the ideal shape.

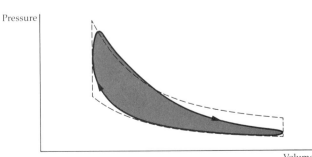

Internal Combustion

The Stirling engine had received little attention until recently because there was an easier way to design small, fairly efficient, compact engines appropriate for mobile applications, first for the automobile, then later for flight. The *internal-combustion engine* took the world by storm. We shall now see how we literally ride on chaos, for the internal-combustion engines of Otto and Diesel are driven by the collapse of energy into incoherence, and abide by the Second Law.

We shall look briefly at both the Otto engine, the basis of the gasoline-powered internal-combustion engine, and the Diesel engine. With neither shall we go into all the technicalities of real engines. Instead, we shall stick to simplified cycles, called the *air-standard* cycles, in which we pretend that the working fluid is air, rather than the awesome mess of gases that actually comes and goes inside a real engine. This will be enough to show the engines going through their paces, and to reveal the principles of their operation.

The cycle for a four-stroke gasoline engine was first proposed by Beau de Rochas in 1862. It is called the *Otto cycle*, because Otto succeeded in making an engine that worked (see below). The actual sequence of steps is shown schematically on the facing page, and the model of the engine in the universe is shown on page 96. The labels A, B, . . . denote the same stages in each illustration.

In the first stage, starting at A, a mixture of air and vaporized fuel is sucked into the cylinder as the piston moves out. This occurs at constant atmospheric pressure (except in unconventional engines), and is represented by the horizontal line running from A to B in the figure below.

The indicator diagram for the idealized Otto cycle. Ignition of the fuel/air mixture occurs at C, and is complete while the piston is stationary.

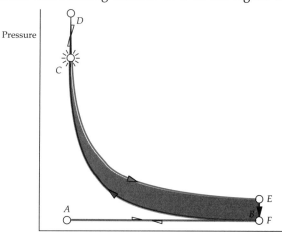

In the next stage, the mixture of air and fuel is compressed as the piston moves in. This step is supposed to be adiabatic, which is a fair approximation to reality; so more and more of the particles in the cylinder are turned ON as their thermal motion is stimulated by the incoming piston. If we want to get as much work as possible from the engine, the pressure should be raised as much as possible (so that the indicator diagram gets as fat as possible). However, since the temperature rises also, there is a danger that the fuel will ignite too soon, which would raise the pressure even more, and we would have to push the vehicle (that is, do work *on* the engine) in order to drive the piston all the way in. For this reason, compression ratios are limited in practice to around 9 or 10.

The Otto cycle. The fuel/air mixture is admitted at A and drawn in as the piston moves to B. It is compressed in the stage from B to C, and ignited at C. The temperature and the pressure rise, and this takes it to D. The power stroke then occurs as the piston is pushed out adiabatically to E. Then the exhaust valve opens, and the gases return to atmospheric pressure and temperature at F. They are pushed out into the street as the piston moves to A.

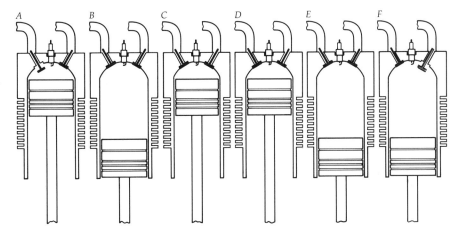

At *C*, the end of the compression stroke, a spark ignites the mixture; its temperature rises as energy pours into the thermal motion of the particles. This energy is released from the chemical bonds that hold the molecules of gasoline together. Many particles are therefore suddenly turned ON: the fuel is burning. The piston is stationary during this rapid ignition: so the pressure rises as the rapidly moving particles strike the walls. This takes the engine to the condition *D* in the illustrations.

Now the crank has turned to the point at which the piston starts to move out. This expansion stage, the power stroke of the engine, is supposed to be adiabatic. That is, the particles thumping against the walls give up their energy to the coherent motion of the atoms of the piston (which can respond by moving) and progressively turn OFF, because their thermal motion is not stoked up again by a supply of energy from outside.

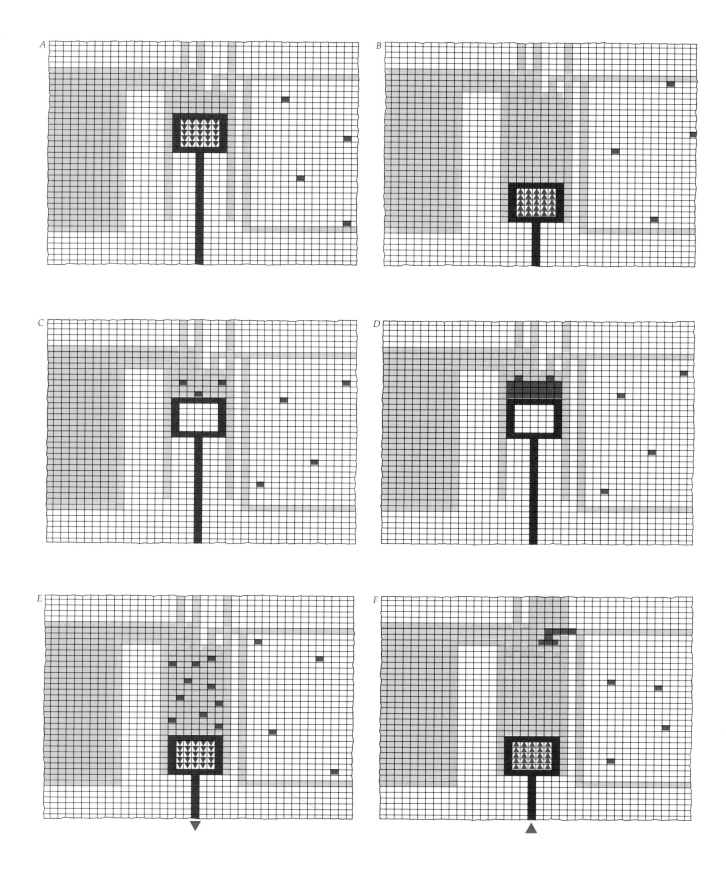

The Mark II model of the Otto engine. The yellow particles in A represent the fuel/air mixture wandering into the cylinder as the piston is withdrawn. This takes the engine to B, when the compression begins, resulting in C. The step is adiabatic, and so some atoms are turned ON. The ignition of the mixture at C causes many of the particles of gas to turn ON, and this brings the engine to D. The violent thermal motion of the particles is transferred as coherent motion of the particles of the piston to the outside world, and in this adiabatic step the particles in the gas turn OFF, which brings it to E. The exhaust valve opens, and its atoms are turned strongly ON by contact with the ON atoms that remain in the gas; the released gas sinks to atmospheric pressure and temperature at F. From F to A is then a spring-cleaning exercise.

At *E* the exhaust valve opens, and the pressure and temperature drop to their atmospheric values (or whatever the local conditions happen to be). At this stage heat is being dumped, as the Second Law demands, into the metal of the engine block (and whatever cooling system it possesses). The exhaust valve takes the brunt of the thermal stress, and is the primary depository of the heat discarded to meet the demands of the Second Law. The valve is the thermal pivot of the engine. Finally, at *F*, the piston moves in, and dumps the cool, atmospheric gases into the street.

At each stage of the Otto cycle, chaos has been the driving force. Energy has tumbled out of chemical bonds; energy has wandered out into the engine block; and the fruits of combustion, such as they are, have dispersed chaotically into the external world. Though coherent motion has been extracted from the engine (and the atoms of the vehicle and its passengers have moved coherently forward), that coherence has arisen from the generation of chaos.

The *Diesel cycle*, proposed by Rudolph Diesel in the hope of achieving motive power from the combustion of powdered coal, is quite similar to the Otto cycle, but differs from it in several important ways. We can follow it through the indicator diagram (below), the schematic representation of the engine on page 98, and the universe model on page 99.

The advantage, in principle, of the Diesel over the Otto engine is apparent from the first stage of the cycle. The retraction of the piston (from *A* to *B*) sucks in air alone, not an air-fuel mixture. The adiabatic compression can therefore take place up to high pressures, and therefore up to high temperatures, because there is no fear that the gas will explode. Only at *C* is fuel sprayed in, and the high temperature it encounters (in other words, the large proportion of atoms that are ON and can bang into its molecules sufficiently vigorously) is enough to ignite it without any electric spark. In

The indicator diagram for the Diesel cycle. Note that ignition occurs as the piston moves out (from C to D).

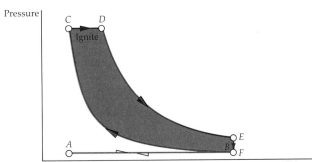

The Diesel engine. Air enters as the piston withdraws from A to B, and is then compressed from B to C. At this point fuel is sprayed in, and ignition of the hot mixture continues as the piston withdraws to D. At D the process becomes adiabatic (all the fuel has been burnt), and the power stroke of the engine takes it to E. There the exhaust valve opens, and acts as a local heat sink for the hot gases, which drop to atmospheric pressure and temperature. They are expelled from the cylinder in the step from F to A.

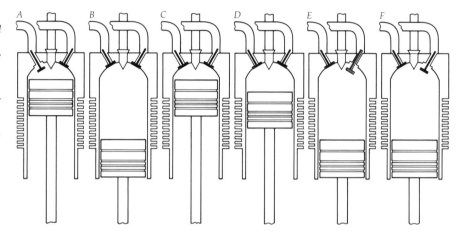

the ideal Diesel cycle, the ignition of the fuel occurs at constant pressure, while the piston is actually moving out; so the temperature rises (because of the combustion of the fuel) at the same time that the volume increases, as the piston is withdrawn. Then, when all the injected shot of fuel is burnt, and while the piston is still moving outward, the temperature and the pressure fall, because now the expansion is adiabatic. At the end of this power stroke, the cycle is at *E*.

At *E* the exhaust valve opens, and the hot, high-pressure gas inside the engine comes to the same temperature and pressure as the nearby outside world, as atoms and energy jostle into dispersal. Once again the metal of the engine is the immediate sink for the heat: the crucial element in the engine is again the exhaust valve, where the chaos is generated that makes the cycle spontaneous and hence the engine useful. (A truck can be thought of as deriving its motive power from the chaos that its engine

The Mark II model of the Diesel cycle. In the stage from A to B, air particles wander in as the piston withdraws; in the stage from B to C some are turned ON. At C the fuel (green) is sprayed in, and enough ON atoms are already present to ignite it. The combustion occurs in the stage from C to D, and many atoms are turned ON by the energy released by the reaction. The incoherent motion of the atoms is picked up as coherent motion by the particles of the piston, and the power stroke takes us to E. Atoms turn OFF during this adiabatic step, but those that are not give up their ON-ness to the particles of the exhaust valve. This takes us to F. At this point the particles wander out into the great outdoors, and are helped on their way by the incoming piston, which takes the engine to A.

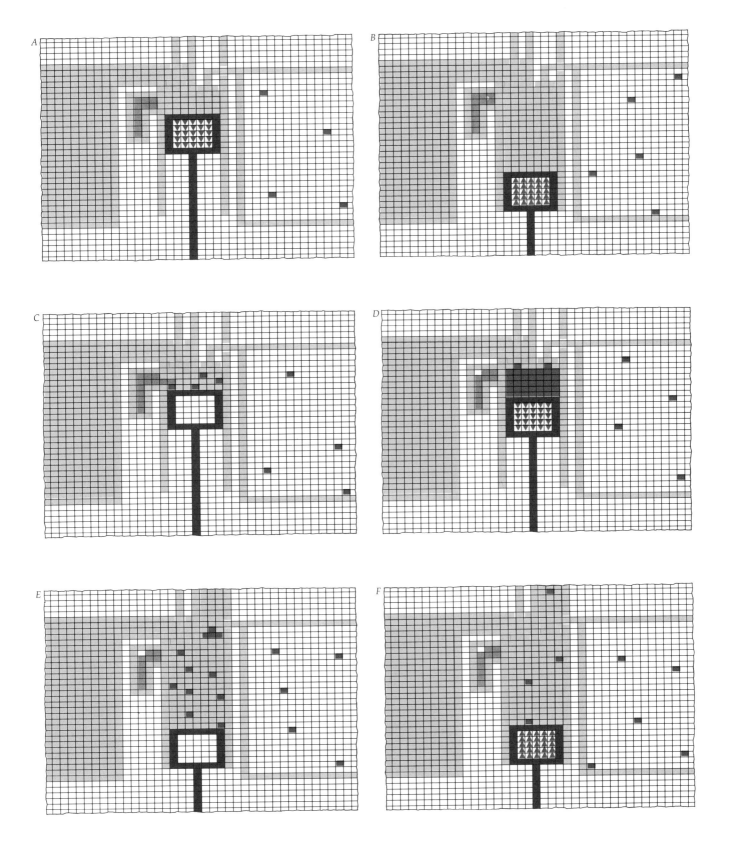

generates in its exhaust valve.) Then the piston starts to come in. As it does so, it drives the gases in the piston out into the street. This brings the cycle back to A. Not only have we generated a little chaos, but we have also moved the vehicle some way along the street.

There are two points worth making before we leave these two every-day engines. First, we have been describing cycles in *four*-stroke engines, in which the crank rotates twice in order to achieve one power stroke. The excursions from A to B and from F to A in the two cycles represent one complete turning of the crank, but contribute nothing to the power output of the device: they are shopping and spring-cleaning expeditions. If we can eliminate them, we *may* be able to improve the efficiency of the engine. In engineering it is a general truth that unnecessary processes are wasteful processes, not merely neutral (the Second Law will always seek out some friction). Greater efficiency *may* result from eliminating them; however, the actual method we contrive to eliminate them may also result in even more unwanted dissipation.

By eliminating the extra turning of the crank, we arrive at the *two-stroke* engines, in which a power stroke falls in each revolution. In the two-stroke Diesel engine, for instance, the cycle is that illustrated below: at A air is *blown* into the cylinder, forcing out the exhaust gases, and replenishing the cylinder with fresh air, readying it for the next cycle. In practice the blower is run from the engine itself. This uses up some of the power output from each cycle, but overall efficiency is increased, because we now have one power stroke in each two-stroke cycle.

The second point concerns another intrinsic inefficiency in the Otto and the Diesel cycles, one that is quite separate from the inerradicable thermodynamic inefficiencies of heat engines. In each cycle the power stroke ends when the piston is at E, that is, when the gas inside is hot and

The indicator diagram for the two-stroke Diesel engine.

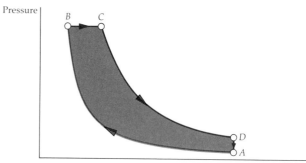

under pressure. Step *E* to *F* in each cycle merely squanders the high-quality energy stored in the gas, and makes no attempt to withdraw it as work. We know that it *is* high quality for several reasons, one being that it tumbles out into the world at such high temperatures, and energy stored at high temperature is of excellent quality. In order to take advantage of this high-quality energy, we should find some way to use the hot exhaust gas, not merely dump it into the universe.

One way of capturing this quality before it is donated as a free gift to the exhaust valve (in either kind of engine) is to attach a *turbine* to the exhaust, and thus lower the gas down to the temperature of the outside world gradually, instead of wastefully squirting it there. A turbine is a device for extracting coherent motion from chaotic motion, just as a reciprocating engine is, but a turbine goes around instead of back and forth; we shall see more of its operation shortly. Turbines are efficient, but normally are limited by the fact that they have to withstand *continuous* high temperatures rather than the periodic surge of temperature characteristic of a reciprocating engine. However, we are now considering using the relatively low-temperature gases at the tail end of the Otto or Diesel engine; so this problem is not severe. We can therefore take advantage of their intrinsic efficiency without having to trouble about metallurgical problems.

The combination of a reciprocating engine and a rotating engine allows the adiabatic expansion (which originally led from *C* to *D*) to continue to *X* (see below). This extracts, without too many deleterious losses, more of the energy released by the combustion of the fuel, and is a commercially effective way of increasing the efficiencies of engines. The turbine need not be connected to the direct load that is being driven by the engine—the wheels of the vehicle, for instance—but may be used to blow air into the cylinders and add to their efficiency that way. This is the idea behind the *turbo-supercharging* of truck and automobile engines.

The indicator diagram for the combination of the Diesel engine with a turbine working from the hot exhaust gases. The turbine cycle is DXAD.

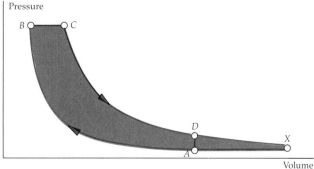

Turbine Power

We are not all confined to the surface of the Earth: chaos can give us wings. In order to complete this rapid survey of the ways in which we have learned to harness chaos and use it to our advantage, we can consider the turbine, which deserves much more than a passing glance, because it is the heart of much modern power. Turbines are anchored to the Earth to generate electric power, and fly through the air in airplanes.

The operation of a turbine can be idealized by yet another cycle, the *Brayton cycle*, which is based on the collection of processes shown schematically below. The compressor stage may be driven by a reciprocating engine, but it is more appropriate to think of it as a rotating fan of some kind. We shall start with the *closed* cycle, in which the same working fluid circulates indefinitely (as in the Carnot and the Stirling engines). Then we shall open the cycle, and allow the engine to take flight.

The closed Brayton cycle runs as follows (see figure on facing page). First, the working fluid is compressed by a compressor, which is driven by bleeding off some work from a later part of the cycle (this bleeding is represented by the line joining the compressor and the turbine in the illustration below). This compression is adiabatic, and it raises both the temperature and the pressure of the gas. It takes the state of the gas from *A* to *B*.

At this stage, *B*, energy is transferred to the high-temperature, high-pressure gas (because fuel burns, or because there is some kind of heat exchanger fueled by a hot source). Its temperature then rises still further, but the engine is arranged so that at the same time the volume of the gas is

The format of a turbine installation for converting heat into work. The hot source (which may be linked to steam boilers) is in red; the cold sink (which might be a nearby river) is in blue. Some of the output of the turbine is used to drive the compressor.

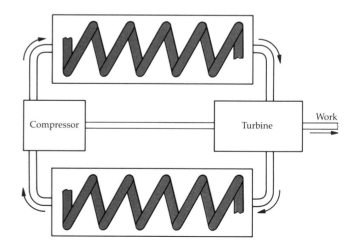

The indicator diagram for the Brayton cycle. The compression and expansion stages (A to B, and C to D, respectively) are both adiabatic.

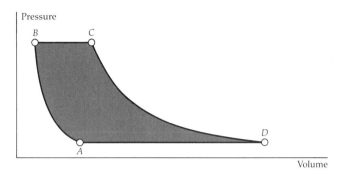

allowed to increase; overall, therefore, it remains at constant pressure. This brings its state to C.

At C the hot, expanded gas enters the turbine. That stage is represented in the cycle by the adiabatic expansion from C to D, which cools it and extracts its energy as work. This time the work results in the coherent motion of the atoms of the blades of the turbine. Finally, in order to return to A and to close the cycle, we must lower the temperature at constant volume. Here once again, we dump heat into a sink in order to complete the cycle and to achieve a viable engine. The technical difficulty of making this cycle practicable has already been mentioned, but is worth emphasizing: the hot and cold devices are separate; so the turbine must be kept at a high temperature while it is running. Turbines therefore became feasible only after metallurgists had developed materials able to withstand high temperatures for long periods.

We can interpret the Brayton cycle in terms of the behavior of individual particles, much as we have already done for the other cycles. Work is done in the stage from C to D, in which the incoherent thermal motion of the hot gas is transformed, at least partially, into the coherent *rotational* motion of the blades. Likewise, in the compression step, work is done as the rotating blades compress the gas: their coherent motion is passed on to the particles they happen to hit. Almost immediately, the coherent motion of the particles is lost as they collide with each other; so their energy becomes stored as thermal motion. The fact that coherent motion is being delivered from and to rotating blades, not reciprocating pistons, is not a significant difference in principle, but it has profound consequences for the smooth and efficient operation of the engine. The particles do not care whether they are hitting the same surface (a piston) as it moves toward or away from them, or a constantly renewed surface (a turbine with many blades).

The compression stage turns many atoms ON in the usual way, and also confines them to a smaller volume. Their natural tendency to disperse, magnified by their higher speeds (higher because their energy is stored in the kinetic energy of their motion), is interpreted by the external observer as an increased pressure. Then, as the fuel is burnt, and its released energy jostles out into the gas (or, if the source of heat is external, as the energy jostles in through the thermally conducting walls), the atoms are stimulated to even greater ON-ness. This takes the gas from B to C. Now the tendency to disperse plays out its role, and the particles enter the region of anisotropy of mechanical response represented by the turbine. From C to D we see the particles losing their ON-ness, not to a single surface, but to a constant succession of surfaces. This reduces the temperature and the pressure, and the engine comes to D. The final stage of the cycle is to relinquish the thermal motion of the gas atoms to their environment, and to reduce the volume as the less-energetic atoms retreat from the constant-pressure surroundings.

Now we *open* the system, and prepare the engine for flight. We replace the *circulation* of the working gas by an open flow. Now the engine has an input and an output (see below); new fluid is constantly drawn in and old fluid is discharged. This results in the *open Brayton cycle*, a model of the jet engines used for flight. The cycle remains thermodynamically similar to the one we have already considered: the incoming gas (the air) is still compressed by the compressor (although the passage of the engine through the air also contributes to the compression). Fuel is still burnt in order to raise the temperature and increase the volume occupied by the passing air. The hot gas still does work on the outside world. In an actual engine the extraction of work occurs in two stages, each of which has a turbine (see below). The smaller of them is used to drive the compressor; the larger is the one the external world sees as the device's muscle. In jet flight, though,

An engine representing the open Brayton cycle. The open cycle is a model of jet propulsion. The left can be thought of as the front of the aircraft, the right as the rear.

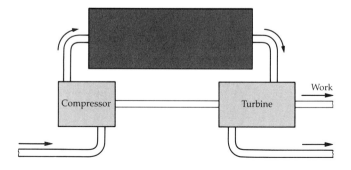

the second turbine is conceptual: the gases simply stream out of the back of the engine. This stage corresponds to work, because the impulse of the particles being thrown out of the rear of the engine is imparted to the aircraft as a whole (as any iceskater who has thrown a ball will have experienced too); so all the atoms of the aircraft and its passengers are moved coherently forward.

Toward Coherence

The central theme of our discussion so far is that chaos can be constructive, and that coherence may stem from incoherence. So far we have seen this in a simplistic way: we have seen that *so long as a process is occurring in which more chaos is generated than is being destroyed, then the balance of the energy may be withdrawn as coherent motion.* We have seen that natural change arises as the Universe slips into, and is trapped in, states of ever-increasing probability. But we have also seen—and this is the crucial point—that the state of greater probability, the state of more chaos, can allow *greater* coherence locally, so long as greater dissipation has occurred elsewhere.

We have seen something of the devices that meet the demand of the Second Law for a cold sink, but in ways that allow us to extract motive power from the uneven way in which the world sinks into chaos. We have seen modifications of the Carnot cycle that describe, at least in theory, the operations of viable engines. Thus we have seen that coherent motion may emerge from incoherent motion if the devices are complex (remember the figure on page 12, the realization of the Brayton cycle). But *mechanical coherence*, the coherence of the motion of particles, is only one aspect of structure. We have seen that achieving coherent motion enables us to build cathedrals (instead of merely heating the stones where they lay) and to move passengers and loads. We may suspect that the same kind of extraction of coherence, more elaborately contrived, may permit us to build bodies. This is the next issue we have to explore.

6 TRANSFORMATIONS OF CHAOS

A new aspect of the degeneration of quality entered surreptitiously toward the end of Chapter 5, and has been lurking in the background. In order to have a hot source we have often either implicitly or explicitly been using *fuels.* That is, to bring about the coherent motion of atoms, we have been using the energy released by certain chemical reactions. The time is therefore ripe to see how the atomic interpretation of the Second Law embraces *chemical* change as well as the physical changes we have been considering so far. We now begin to see the structural potency of chaos, and recognize that collapse into chaos can transform materials, and even bring bodies to life.

Cooling is one of the simplest kinds of physical change: we have seen that the natural cooling of a hot body to the temperature of its environment can be readily accounted for by the jostling, purposeless wandering of atoms and quanta that we call the dispersal of energy. We shall now see that chemical reactions, processes in which one substance changes into another, are no more than elaborate forms of cooling. In cooling, the atoms of some substance give up their thermal motion by frittering it away chaotically into the environment and undergo no other change. In a chemical reaction, energy is dissipated, but atoms also change their partners. As a result, a new material is formed as the original substance (or substances) changes into something different. All chemical reactions are elaborations of cooling, even those that power the body and the brain. Consciousness itself is a consequence of the gradual cooling of components of the Universe.

Before we can fully appreciate this last remark, we need to fill in some background. As a first step we shall explore what goes on in the course of a simple chemical reaction, and see what it is that decides which reactions may occur, and which ones may not. We shall have to distinguish the *tendency* of a reaction to occur from the *rate* at which it does so. For instance, although hydrogen and oxygen tend to react with each other (because in that way the Universe cools), a mixture of these two gases can be kept unchanged for years unless we stimulate the reaction with a spark. Then, having made the distinction between tendency and rate, we shall go on to see that both the direction in which spontaneous chemical reactions go and the rates at which the reactions take place are governed by the dispersal of energy, but in different ways.

In the course of this chapter, therefore, we shall see how an elaboration of the process of cooling brings about the transformation of matter. We shall also see an analog of the discussion of Chapter 5, where chaos brought out coherence from incoherence as heat was transformed locally into work: some materials are constructed as substances give rise to more ordered products. But here as there, such apparent reversals of the stream of Nature are *local*, and are driven by the generation of greater disorder elsewhere.

Chemical Transformations

Initially we shall concentrate on a single chemical reaction, the burning of iron. The burning of iron might not appear to be a very common chemical reaction: coal, yes; iron, no. However, it is in fact one of the most important and common reactions in the world; we usually call it *rusting*. Moreover, the burning of iron is also the primary step in the sequence of processes that allowed humanity to develop and to master its environment, because *respiration* also begins with a step that corresponds to the combustion of iron. In that step the oxygen of the air is attached to the iron atoms in the hemoglobin molecules in the red corpuscles of our blood. That blood is rust-colored is no coincidence: respiration is a kind of rusting. We could develop an engine that used iron as a fuel for combustion (an iron locomotive could run by consuming itself), but Nature has beaten us to it: the processes of our bodies are at least partly fueled in this way.

In order to see what form the cooling of the Universe takes when iron reacts with oxygen, we need to know something about chemical bonds. A *chemical bond* is the link between atoms that holds them together in the definite arrangement that distinguishes one type of molecule from another. The fundamental reason for the link between two atoms is to be found in the reduction of energy that occurs as a bond is formed: if the energy of the molecule is less than the energy of the separated atoms, then the bond is stable and the molecule survives (see the top figure on the facing page). The contributions to the energy are numerous and subtle, and the stability of bonds results from many quantum-mechanical effects (but we do not need to know their details here). For simplicity, we can regard the stability as being due to the lowering of energy as the negatively charged electrons and the positively charged nuclei of the atoms adjust their relative positions and move to energetically more favorable locations.

A stable bond is formed if the energy of the resulting molecules (on the right) is lower than the total energy of the original particles (the atom and molecule on the left).

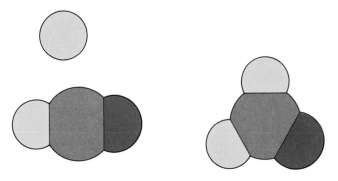

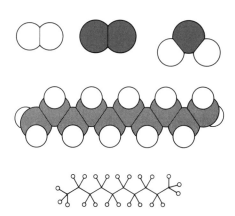

Molecules come in all shapes and sizes. Four simple examples are shown here. The smallest is hydrogen (H_2), represented by two white atoms. Oxygen occurs as the diatomic molecule O_2 (the pair of red atoms). Water is the triatomic combination H—O—H (abbreviated to H_2O). The long zigzag chain is the hydrocarbon $CH_3CH_2CH_2CH_2CH_2CH_2CH_2CH_2CH_2CH_3$, or more briefly $C_{10}H_{22}$, known as decane (a component of oil). The carbon atoms are shown in dark gray. We shall use this color code throughout.

Molecules come in all shapes and sizes. Each different material has molecules whose atoms are grouped in a characteristic arrangement. The simplest molecule of all is the *hydrogen molecule*. It is *diatomic*, that is, consists of *two* hydrogen atoms, which stick together with their nuclei separated by about 7.5×10^{-11} meters. That distance is called the *bond length* of the molecule. A molecule of oxygen has a similar structure: it is also diatomic, and the nuclei of the two oxygen atoms are separated by about 1.2×10^{-10} meters. We can think of these molecules as depicted on the left: the oxygen molecule (O_2) is bigger than the hydrogen molecule (H_2) because it contains 16 electrons, but the hydrogen molecule has only two.

Whereas a hydrogen molecule is held together by the electrostatic interactions of only two electrons and two nuclei, a lump of iron results from the interactions of myriads of atoms. A lump of iron (or any metal) can be thought of as an array of nuclei; some of the many electrons present roam through the array of nuclei (see the figure at the top of the next page) and act as an all-pervading electrostatic glue. Only some electrons are free to roam, for the strong charge of the nuclei of iron atoms grips many of the electrons tightly, so that they cannot escape from their parent atom. We can depict a lump of iron as a stack of almost spherical iron *ions** surrounded by a sea of the electrons that do manage to escape from each atom. This sea is called the *Fermi sea*. Even though each atom donates only a few electrons to the sea, a kilogram of iron contains more than 10^{25} atoms; so the total number of electrons in the sea is huge.

---

* An ion is an atom that has an electric charge because it has gained or lost one or more of the electrons, which have a negative charge. The iron ions have *lost* electrons.

Metallic iron can be thought of as a collection of positively charged ions (the blue spheres) surrounded by a sea of electrons, which can migrate freely through the lattice.

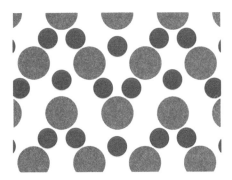

This is a schematic representation of the iron-oxide lattice. The positively charged iron ions (blue) and the negatively charged oxide ions (red) are held rigidly by their electrostatic interaction.

Oxygen is a gas composed of discrete O_2 molecules, a swarm of tiny particles, each with its puddle of electrons. Iron is a metal, a great stack of ions stuck together by the electron sea. The electrons of that sea are mobile, and can readily be moved from place to place. This gives rise to the characteristic properties of iron and other metals, such as their electrical conductivities and their lusters. It also gives rise to iron's malleability, since groups of ions may be pushed past each other through the sea; so an ingot may be hammered into an automobile or some other artifact of civilization.

So much for iron and oxygen. The only other substance we need consider at this stage is iron's oxide which we commonly call rust; this combination of iron and oxygen is the ash from their burning. This ash is a great pile of iron ions and *oxide* ions, the latter being oxygen atoms that have gained a couple of electrons, and so are now negatively charged.* The two kinds of ions are held rigidly together by the attraction between their opposite charges (see figure on left).

The point to filter from this discussion is that when atoms group together and give rise to various materials, different amounts of energy are stored away. The energies locked up in a collection of oxygen molecules, a lump of iron, and a pile of iron oxide are all different, just as the amount of energy locked up in a hot lump of iron is different from the amount stored

---

* To be precise, we are considering the product of the combustion to be iron(III) oxide, which is an ionic solid formed from the ions Fe^{3+} and O^{2-}. The combustion reaction we shall have in mind throughout the course of the following discussion is

$$4Fe(s) + 3O_2(g) \rightarrow 2Fe_2O_3(s).$$

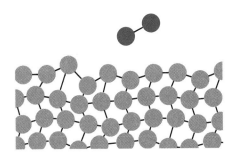

The initial scene in the sequence of events involved in the oxidation of iron. At this stage the iron ions of the metal are vibrating, and their distances from their neighbors are fluctuating. Above them fly vibrating, rotating oxygen molecules.

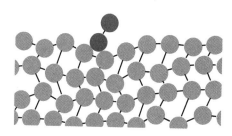

The second scene. An oxygen molecule strikes the surface: the heavier the landing, the better for what follows.

when it is cold. *That* is why reacting and cooling are similar: different energies are stored in different substances, and in different states of the same substance; and physical and chemical change involve changes in the quantity and manner in which that energy is stored.

Iron Burning

All chemical reactions are complicated affairs, but we can see the *principles* behind them by simplifying what actually happens. Therefore what we are about to describe is a *model* of the reaction process, which is intended to focus attention on the kinds of things that go on: it is not a description of the precise sequence of events in the turmoil of incandescence (or, for actual rusting, the slow smoldering which for an individual atom amounts to a brief incandescence) that accompanies the burning of an actual piece of iron. The model captures the essence, not the detail.

We model the reaction as follows. The particles of the lump of metal are vibrating vigorously (upper figure on left), which means that the distances between them are fluctuating: at one instant the nucleus of one ion may be uncharacteristically distant from its neighbors; at the next instant the distance between them may be uncharacteristically short. All the atoms are swinging around in the manner typical of thermal motion. If the temperature is high, the motion is vigorous, and the excursions of an ion from its neighbors might be large. The oxygen molecules are all zooming around with the chaotic motion typical of particles in gases. They are also vibrating: the two atoms of each molecule are swinging toward and away from each other, and the bond length is periodically lengthening and shortening. This vibrational motion is another store for energy, and its vigor increases as the temperature of the gas is raised.

The oxygen molecules strike the surface of the iron (lower figure on left). We shall follow the fate of one of them. If the block is hot enough, and the collision sufficiently vigorous, some of the atoms will happen to be in an arrangement that includes (more or less simultaneously) a long oxygen bond, a short iron-oxygen distance, and a long distance from that iron ion to its neighbors (upper figure on next page). This can be recognized as the *incipient* formation of a bond between the iron and the oxygen, and the *incipient* destruction of the original iron and oxygen bonds. This incipient rearrangement is accompanied by a release of energy (so long as the new bond is stronger than the old). The released energy is picked up by the neighboring particles, which burst into vibration (more are turned ON). As a result of their impacts with their neighbors, this energy very quickly jostles away (lower figure on next page).

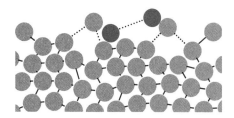

While the oxygen is in contact with the surface, the distances of the atoms in the vicinity may undergo large fluctuations (the chance of this happening depends on the energy available). In this scene, we see two oxygen atoms incipiently losing each other and forming close associations with similarly released iron ions.

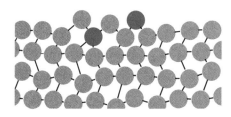

In this final scene, the energy released by the reorganization of bonds has jostled away, and the iron and oxide ions have formed permanent associations. Iron oxide has begun to form.

Because the energy disperses from the site of action, the atoms are trapped in their new arrangement. In particular, we have the beginnings of a lump of iron oxide, in which an ion of iron has been ripped from its neighbors and an oxygen molecule has been torn in half. The new entity cannot wobble back into the arrangement that the reactants had earlier, because doing so would require energy to reassemble spontaneously. Just as in ordinary cooling, the chance of energy reassembling by itself is so remote that we can regard the iron-oxygen bond as locked into perpetuity. Boltzmann's Demon can go on reorganizing and rearranging the energy, but it is extremely unlikely that it will ever stumble on the arrangement that accumulates energy at the point where it would liberate the iron and the oxygen from their bond. The iron has burnt, and it will stay burnt.

The *staying* burnt of the iron is the important feature. The reaction is irreversible, like the cooling of hot iron. The Universe has stumbled into a state of higher probability as a result of the wanderings of the reactants and the liberation of their energy. Once there, the Universe is trapped. It cannot spontaneously return to a state of lower probability: the Demon may rearrange for ever, but few rearrangements lead to the recovery of the reactants; its stumbling onto the arrangement that would destroy the bond is so unlikely that we can, for all practical purposes, consider it impossible.

There is also another important point. The products of the reaction have less energy than the reactants. The energy in the bonds binding the product is less than was in the bonds binding the reactants: the excess has been carried away in thermal motion. The reaction has proceeded downward in energy of the substances. The tendency of the reactants to change thus resembles the tendency a ball might have when left on the side of a hill. However, the analogy is as misleading as it is close. The direction of change is not *directly* related to the quantity of energy stored by the bonds. Although the products possess less energy than the reactants, so that the substances involved have dropped to a lower energy, this is not the *reason* why the reaction has taken place. Overall the energy of the Universe remains constant: the *Universe* has not dropped to a lower energy in the course of the reaction. All that has happened is that some initially localized energy has dispersed. *That is the cause of chemical change: in chemistry as in physics, the driving force of natural change is the chaotic, purposeless, undirected dispersal of energy.*

But has the *dispersal* in fact increased? Here we can identify a complication. It reveals some of the complexities we encounter when we deal with chemical (rather than physical) change. Let's look more closely.

In the reaction a small lump of iron has reacted with several liters of oxygen; a kilogram of iron in fact requires around 300 liters of oxygen for its

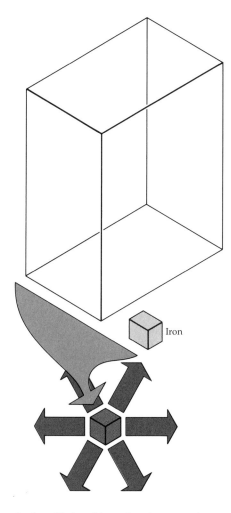

In the oxidation of iron, there is a competition between modes of energy dispersal. Although energy is released and spreads (the red arrows), the original large volume of gas (much larger in proportion than is shown here) is in effect confined to a small pile of oxide.

complete combustion under atmospheric conditions. The product is a little pile of oxide. Initially, the energy of the oxygen molecules was very widely dispersed; after the reaction all the oxygen atoms are in that little pile of product. Overall, therefore, although energy is liberated by the reorganization of the bonds, there is also a *localization* of energy, because 300 or so liters of oxygen have been confined into about a liter of rust.

Whether or not energy is dispersed overall in the reaction depends on the outcome of a *competition* (see figure on left). On the one hand there is the localization of energy as the gas collapses into the product; on the other there is the dispersal of energy as a result of the reorganization of the bonds. The only way to resolve the conflict is to attach numbers to the varieties of chaos that are coming and going as the reaction proceeds.

In order to attach numbers, we use the information set out in the earlier chapters. The following is the *principle* behind what is needed, not the precise details.

One contribution to the entropy is the change that occurs because energy is liberated into the surroundings by the rearrangement of the bonds. Since this change is given by *(Heat supplied)/Temperature,* and since it is easy to measure the energy given out as heat when a reaction occurs (especially if it is a combustion), this contribution to the chaos can be calculated very easily. The entropy change is positive, because energy is dispersed into the surroundings when it becomes available as the bonds rearrange.

The other contribution is the change of entropy that occurs as the substances themselves undergo chemical change: the oxygen is eliminated, and its atoms combine with the iron. Not only do we lose the iron metal, but we also gain a complex ionic solid. Although the changes that have occurred are complicated, and the energies of the iron, the oxygen, and the oxide are stored in complex ways, the entropies of each individual substance can be *measured.* (We explained how on p. 35: it involves observing the temperature of a sample as it is heated, and manipulating the data.) Therefore, all we have to do in order to assess the change of entropy that arises from changing reactant molecules into product molecules is to find the values for the reactants and for the products, and take the difference.

When this is done, it turns out that the total entropy of the reaction substances *decreases* in the course of the reaction, partly because the oxygen is confined into its iron cage as the oxide forms, and partly because the amounts of energy stored in each type of bond are modified. Not only are the products less physically dispersed than the reactants, but they also possess less energy. Consequently they are less disordered (in the general sense), and therefore have lower entropy too.

For this reaction, the reduction of entropy of the substances taking part (the *system,* in our former language) turns out to be much less than the increase of the entropy in the surroundings. In fact, the latter is about ten times greater, because the reaction releases so much energy (the iron-oxygen bonds that form are exceptionally strong) and creates a lot of disorder in the surroundings. Therefore, *overall* there is an increase in the chaos of the world when iron and oxygen make way for rust. Hence steel artifacts are *intrinsically* unstable, and automobiles have an ineluctable tendency to smolder to death.

Cooling As Heating

We have seen that a more structured (lower-entropy) product may emerge from less-structured (higher-entropy) reactants if compensating chaos is generated in the surroundings. This is the analog of heat producing work. That is the first stage in our examination of the manifestations of chemical change. We shall return to it later, but first there is another aspect to explore, one which emphasizes the wide range of the deceitful but welcome appearances that Nature can adopt as it disperses in chaos. We shall now see that chemical cooling may take a particularly bizarre form: it can correspond to the *inflow* of heat.

In order to see that this is so, we now consider a second kind of reaction. A simple version of a general class is one in which a molecule falls apart into two fragments (see top figure on facing page). This could be denoted A—A $\longrightarrow$ 2A; we shall suppose that all the components of the reaction, the reactant A—A (A_2 for brevity) and the product A, are gases.* We shall also arrange for the pressure and the temperature of the sample to be held constant throughout the reaction.

We can see what must happen for the reaction to take place. First, energy must be available to break the A—A bond. The bond may be weak, but the atoms don't just fall apart. In contrast to the iron combustion, no new bonds are formed; so there is no compensating lowering of energy. There must therefore be a net *supply* of energy to the sample if the reaction is to proceed from pure A_2 toward A. Therefore the products, if they form, lie at a *higher* energy than the reactants. There is a naive view that reactions

---

* An actual example of this reaction is the decomposition of dinitrogen tetroxide (N_2O_4) into nitrogen dioxide (NO_2) according to the reaction

$$O_2N\text{—}NO_2 \longrightarrow 2NO_2.$$

The reactant is a colorless gas; the product is a dark-brown gas; it is therefore easy to follow the course of the reaction by monitoring the color of the sample.

An example of an $A_2 \longrightarrow 2A$ reaction (called a dissociation): dinitrogen tetroxide, N_2O_4, falls apart at its weak N—N bond to form two nitrogen dioxide, NO_2, molecules. Blue is the color code for nitrogen.

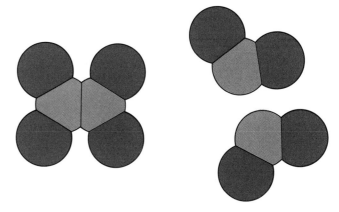

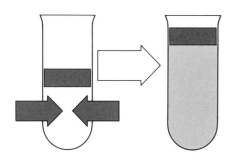

In the dissociation reaction at constant pressure, there is an influx *of energy. We shall see, however, that this still corresponds to a dispersal.*

occur because the reactants roll down to lower energy, like a ball rolling down hill. If this naive view were true, this reaction would not take place, because it corresponds to rolling uphill. Yet examples of the $A_2 \longrightarrow 2A$ type of reaction *are* known. Hence this naive view cannot be tenable.

The reason why a reaction occurs, as we have stressed before, is that it corresponds not to a lowering of energy, but to a *dispersal* of energy: *reactions are degradations of quality, not diminutions of quantity.* We have to track down the various channels of dispersal and identify their contributions to chaos. We shall see that, although energy may indeed flow in (as in figure on left), by doing so it sometimes manages to become *more* dispersed! Therefore, in the sense that cooling corresponds to dispersal, accumulation of energy as the reaction proceeds may also correspond to cooling!

The contributions to chaos in the $A_2 \longrightarrow 2A$ reaction are as follows. There is a reduction in the dispersal of energy, since thermal motion in the surroundings is extinguished when it wanders into the sample and is used to fracture the A—A bonds (see top figure on next page). This corresponds to a reduction of the entropy of the surroundings. When a bond breaks in the sample, wherever one particle was before, now there are two. Unless there are other changes, the reaction causes twice as many particles to be crammed into the same volume as before. As a result, the pressure doubles. (The pressure of a gas, we have remarked, arises from the impacts of particles; now that we have twice as many particles in the same volume, the pressure is twice what it was before.) Therefore, in order to maintain the same pressure, we must allow the volume of the sample to expand to twice its previous value (see bottom figure on next page). This corresponds to a physical *dispersal* of the products (and the energy they carry), and therefore to an *increase* in their entropy.

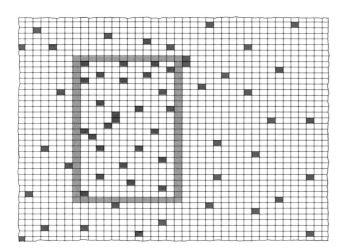

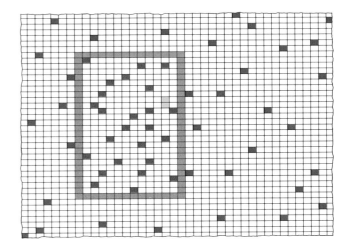

The Mark II universe model of the dissociation reaction. In the course of the dissociation, there is an energy influx. The blue rectangles are A_2 molecules: in order for them to dissociate into two A molecules (yellow rectangles), energy must enter from the environment. This reduces the entropy of the surroundings.

At this stage we can identify three contributions to the change in entropy of the Universe when A_2 molecules fall apart under conditions of constant temperature and pressure. First, there is a change of entropy that results from replacing an A_2 molecule by two A molecules. The A_2 and A molecules store their energies in different ways, and just as the entropy of solid iron differs from the entropy of iron oxide, so the entropies of these molecules differ (and may be measured). The change when an A_2 molecule

In order for pressure to be constant (a condition we are imposing on the reaction), the volume of the container must increase. Compare the initial state (with 26 A_2 molecules) and the completely dissociated state shown here (with 52 A molecules). The volume must be doubled to hold the pressure constant (the temperature is constant too).

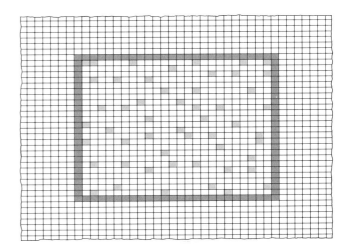

turns into two A molecules may be either positive or negative, depending on the specific chemical compounds involved. It may be assessed from the measured (and often tabulated) entropies of the individual substances by taking the difference between their values.

The second contribution is the change of entropy arising from the volumes occupied by the reactants and the products. It is *positive* in this $A_2 \longrightarrow 2A$ reaction, because the products occupy more room than the reactants.

The third contribution is the change of entropy of the surroundings. For an $A_2 \longrightarrow 2A$ reaction, since energy has to be drawn in to break the bond, this change has to be *negative,* and the entropy of the surroundings is lowered.

The overall change of entropy may be either positive or negative, depending on the relative magnitudes of the three changes. For the specific reaction we have in mind (see the footnote on p. 114), the overall change turns out to be marginally *negative.* That is, for an *actual* reaction of this kind, one that is known to run in the direction of 2A, the changes we have identified so far lead us to expect the entropy to *decrease* as the reaction proceeds (see below). Something is out of joint: we appear to have uncovered a reaction which moves spontaneously forward, but simultaneously lowers the entropy of the Universe!

We must have missed a contribution to the entropy. We have calculated the change for a *pure* substance A_2 evolving into a *pure* sample of A. At no stage have we mentioned the role of *mixing.* Mixing is a form of muddle, and when several species are present in the *same* volume (for

The dependence of the entropies of the system, the surroundings, and their total as pure A_2 transforms into pure A. Notice that the total entropy does not go through a maximum except at the starting point; so this appears to predict that A_2 will remain completely intact.

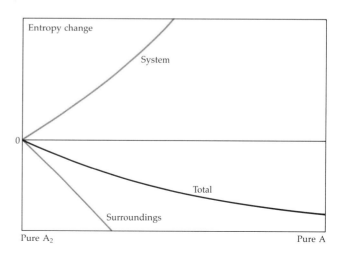

We have neglected the entropy of mixing, and have implicitly assumed that at each stage of the reaction the two components remain unmixed, as depicted here.

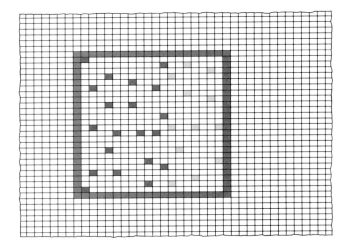

instance, A_2 and A at an intermediate stage of the reaction), the entropy is greater than when only one is present. That, after all, is why gases mix: the spontaneous direction of change is from separate compounds towards a mixture (see above), which corresponds to the increase of universal chaos and therefore also to the increase of the entropy of the world. For the reaction $A_2 \longrightarrow 2A$, there is a mixture present as soon as the first A_2 molecule has fallen apart (see below).

The entropy increases, and the chaos too, when the particles of the two gases mingle. This is why gases mix spontaneously. We have to take this into account in the entropy calculation for the dissociation reaction. The magnitude of the contribution depends on the relative amounts of the species present.

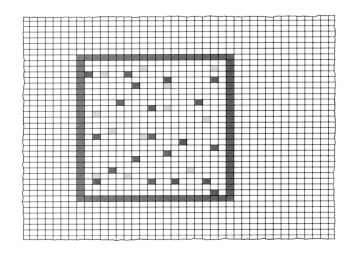

At first, when only pure A_2 is present, there is no mixing, and hence no chaos that arises from mixing. If *all* the reactant molecules were to fall apart, there would also be no chaos arising from mixing, because then there would be only pure A. But at *intermediate* stages, there is chaos that arises from the mixing, because then A_2 and A species are both present in the same container. This chaos corresponds to a *positive* contribution to the entropy, which rises to a maximum when A_2 and A species are present in similar abundances (see below).

The entropy of mixing of two gases at each stage of the A_2 dissociation reaction. The contribution is zero when only one type of molecule is present (left and right), and is a maximum at an intermediate composition. (The horizontal axis denotes the mole fraction *of A present in the mixture.)*

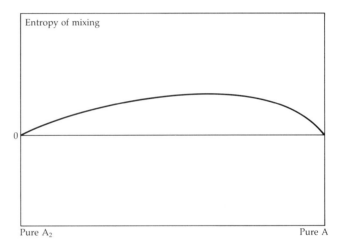

The total entropy change occurring in the course of the reaction $A_2 \longrightarrow 2A$ is now seen to result from *four*, not three, contributions. There are contributions from the changes of the molecules themselves, from the change of volume of the reaction system, from the change in the surroundings on account of the depletion of their energy, and from the effect at intermediate stages of having a mixture. The last contribution adds a hump to the graph at the bottom of page 117, and we obtain the figure on the next page. This shows a small but definite maximum at an intermediate composition.

We can now see what happens in the course of the reaction, and how it "decides" where it comes to rest. If we have pure A_2 initially, then the chaos of the Universe increases if the reactants fall apart to some extent (and an important aspect of this increase of entropy is the entropy of mixing). Reactions like this are therefore driven largely by the muddle that arises from mixing. Not all the A_2 particles fall apart, for that would be going too far and would not correspond to an overall increase in chaos.

The three contributions to the entropy of the dissociation reaction. The blue line shows the increase of entropy of the system, and the brown the reduction of entropy of the surroundings (as in the figure on page 117). The black line is the contribution of the entropy of mixing (see the figure on page 119). The total is shown by the red line: there is a clear (but not very sharp) maximum, which corresponds to the equilibrium composition of the reaction mixture. The mixture tends toward this composition whatever the starting composition.

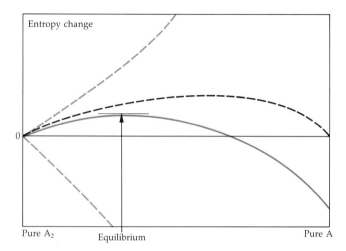

Instead, the reaction proceeds until a *proportion* has fallen apart (exactly how much depends on the conditions of temperature and pressure). The system is then in dynamic equilibrium. This is a high-entropy, highly probable state, and the Universe is trapped in it. The Demon can now deploy not only the energy at its disposal (which it can do in the surroundings less freely than before, because more atoms are now OFF), but also the particles themselves. It can now deploy the particles physically, not only over the greater volume accessible to the system, but also by taking advantage of the intermingling of the different kinds of molecules in the mixture. Only by the remotest chance will the deployments of the Demon reassemble the pure A_2 in its original abundance and volume. The change from pure A_2 to some A_2 together with its product A is, for all practical purposes, an irreversible change.

A similar argument applies when, instead of starting with pure A_2, we take a sample of pure A and inject it into a container. Now we know that the direction of increasing chaos is for some of the A molecules to *stick together* to form A_2. Chaos now appears to be running in reverse, but in fact the Universe is now climbing up the righthand hill of entropy (in the figure above). Now, even though composite structures are being formed out of fragments, the chaos of the world is increasing. The mixture evolves to the same intermediate composition as before, the composition corresponding to the maximum of the entropy hill, the maximum of Universal chaos.

We should now take stock of where we are. We have seen that some reactions may generate chaos by allowing energy to escape into the world, but others may do so by allowing it to wander in, and using it to generate more disorder *within* the sample than has been eliminated outside. The natural direction of any kind of reaction is in the direction of overall chaos. In this sense reactions are akin to the process of cooling, but the deployment of energy that corresponds to dispersal is more subtle. Being more subtle, it might trick the unsuspecting into thinking that a different mechanism is at work.

The Rate of Dispersal

We emphasized earlier that it is important to distinguish the *direction* of spontaneous change from the *rate* at which it is achieved. Spontaneous does not mean fast, although some spontaneous processes are fast. Spontaneous means natural, and refers to a process that can happen without our doing any work to bring it about. The fact that automobiles survive for at least a few years even though they are exposed to fiercely reactive oxygen shows that there is more to chemical change than direction alone.

In another context I have remarked that "chaos is both the carrot and the cart", (see figure below). We shall now see what lies beneath this remark, by showing that the dispersal of energy helps determine not only the direction of natural change, but also the rate at which a reaction proceeds to its destination. In other words, I shall now argue that chaos determines not only destiny but also the rate at which that destiny is achieved.

In order to see how chaos may restrain as well as promote, we shall consider the first reaction again, the oxidation of iron. In particular, we focus on an essential step in the way it was modeled: the vibration of the iron ions to uncharacteristic distances from their neighbors (see top figure on page 112). If this happens when an oxygen molecule is colliding with

Chaos is both the carrot and the cart. It determines both the direction of spontaneous change and the rate at which equilibrium is attained.

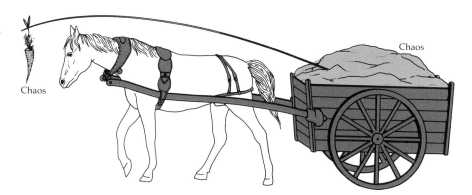

that region of the surface of the metal, then the atoms may discover themselves in an arrangement corresponding to the incipient formation of the oxide. Then, as energy is released into the surroundings and slips away, that new arrangement ceases being merely incipient and is frozen into permanence. The rate of the reaction clearly depends on how often the iron atoms vibrate to large amplitudes. The products are arrangements of atoms stumbled into by misadventure and then preserved by the dispersal of energy; the rate of the spontaneous change depends on the extent to which atoms can explore out to risky bond lengths.

The oxidation reaction depends on the ions' ability to wobble at their sites especially vigorously. But vigor means energy. Therefore, the untypically large and potentially risky vibrations in some location means that an unusual amount of thermal motion is taking place there. That in turn implies a dense accumulation of energy. The Demon's deploying, although it achieves uniformity in general, may also result in these brief accumulations. The rate of the reaction depends on the chance that these accumulations will be assembled; by its effect on this chance, chaos governs rates as well as direction.

The energy that must accumulate in order for a particle to undergo a reaction is called the *activation energy*. The readiness with which this energy is attained locally depends on the temperature. This is most clearly seen in terms of the simple model universe (on the right), because the likelihood that a large number of ONs will accumulate near a given atom depends on the proportion of atoms already ON. If a high proportion of atoms are ON, then accumulation is likely to take place very often, and an oxygen is likely to make successful encounters. If the temperature is low, and only a small proportion of the atoms are ON, then the dealings of the Demon will lead only extremely infrequently to sufficient accumulations, and the oxygen molecules will normally bounce off the surface unchanged.

The likelihood that the activation energy can be accumulated at a particular temperature is given by another expression devised by Boltzmann, the *Boltzmann probability*.* It leads us to expect rates of reactions to increase

---

* The more common name is the *Boltzmann distribution*. It is an expression of the form

$$Probability = e^{-(Activation\ energy/Temperature)},$$

where e is the exponential e. The general form of this expression was deduced by Boltzmann; it was used to describe the rates of chemical reactions by the Swedish chemist Svante Arrhenius, and his name is normally attached to it in this connection. In his doctoral thesis Arrhenius put forward ideas on the nature of matter which generated so much disbelief that he was awarded a pass of lowest rank; he was later awarded the Nobel Prize for them.

At low temperatures, when only a small proportion of atoms are ON, there is only a small
likelihood that enough energy will accumulate at
a molecule for a reaction to be able to occur. At
high temperatures, large-enough accumulations
are likely to occur frequently; so the reaction
may proceed rapidly.

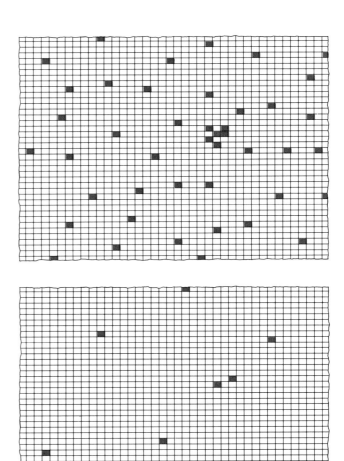

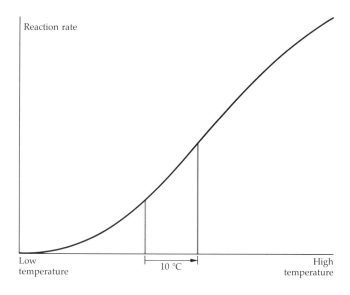

The temperature dependence of the rate of reaction expected from the Boltzmann probability. For typical activation energies, in the vicinity of room temperature, reaction rates double for an increase of around 10 °C.

Reaction rate

Low
temperature

10 °C

High
temperature

sharply with increasing temperature (see above), and this is generally true. For typical activation energies, increasing the temperature by ten degrees doubles the rate of reaction. Tropical flashing butterflies, for instance, flash more quickly on hot nights than on cool.

Chaos and Order

In chemistry, as in physics, the process of change is the outcome of the purposeless operations of chaos. Now we have seen the two faces of Boltzmann's contribution: the manner in which chaos governs direction, and the manner in which it governs rate. We have seen both the carrot and the cart. In particular, we have seen that the Demon's unconscious, purposeless deployings, and the sinking and then trapping of the world in states of ever-greater probability, account not only for simple physical change, as in the cooling of a lump of metal, but also for subtle change, as in the transformations of matter. We have seen that chaos can lead to order. For physical change, that order takes the form of doing work, which can build structures, sometimes on colossal scales. For chemical change, order is also generated out of chaos; but now the order, the definite arrangement of atoms, is on a microscopic scale. Order on any scale can arise from collapse into chaos: order springs locally from disorder elsewhere. Such is the spring of change.

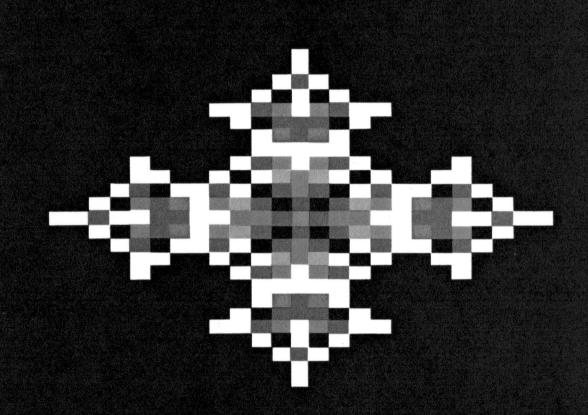

7 POWERS OF TEMPERATURE

Both the direction of natural change and the rate at which it is achieved are aspects of the *distribution* of energy. The direction of natural change depends on the tendency of energy to disperse; the rate depends on the abundance of local accumulations that loosen the grip of atom on atom. Random arrangements of atoms can be locked into a new structure if they release energy that wanders away and leaves them immobile.

Recognizing this dual role of chaos stimulates two thoughts. The first is about how the properties of the real world depend on the temperature, and in particular how they respond to massive changes that take it to the extremes of hot and cold. The second is about how such extremes, especially the extreme of cold, may be achieved. We need to deal with extreme cold because it seems to oppose the Second Law: how can an object be cooled *below* the temperature of its environment? The *natural* direction of change is in the opposite direction; in that sense, cooling is a process that runs against Nature.

The journey that reveals the powers of temperature can start at an everyday scene (see top figure on next page). We may then move higher and lower in temperature in tenfold leaps, from ordinary temperatures, where we can picnic, to ten times hotter, then ten times hotter still; and we can also go to ten times colder, then ten times colder still, and so on. This starting scene for our journey, as readers might recognize, is the pivotal scene of *Powers of Ten,* a companion in this series: there the journey was in powers of ten of distance, here in powers of ten of temperature.

In fact, there is a deeper link between these two volumes. *Powers of Ten* can be regarded as a journey not through scales of space, but through *time,* where the observer on a spacecraft plunging toward Earth inspects the scene at intervals that decrease in powers of ten. Our journey here is in temperature, not time; yet temperature and time as we have begun to glimpse, have some odd relationships. Suppose we represent temperature and time on two perpendicular axes on a graph. The behavior of matter as time and temperature are changed can then be indicated by the motion of a point on the plane that the axes define (see bottom figure on next page). Moreover, if we were to think in terms of *complex numbers* (these are num-

A picnic scene: at these temperatures (around 300 K) a wide variety of chemical reactions take place, and the world is a rich and complex place.

bers involving *i*, the square root of −1, and having the form *a* + *ib*), then *temperature can be regarded as imaginary time.* That is, certain equations in ordinary dynamics are turned into *thermo*dynamic expressions if we replace the time (which is represented by a *real* number, such as 5 or 320) by an imaginary number (a number of the form *ib*, such as 5*i* or 320*i*). We can therefore think of our journey to high and low temperatures away from the picnic as another journey in time, but along the *imaginary* time axis. Just as *Powers of Ten* inspects the world after intervals of real time that increase in tenfold steps, so we inspect the world at tenfold intervals of imaginary time. Together with *Powers of Ten*, powers of temperature span the complex plane of time, and embrace both dynamics and thermodynamics. In the sense of this discussion, we live in the complex plane of time.

Certain technical reasons justify our regarding temperature as "imaginary time"; so we can depict events as occurring at points in the complex plane of time.

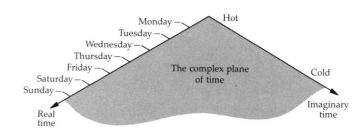

Normal Life

At normal temperatures we can picnic (as in the figure to the left): at typical everyday temperatures (around 20 °C or so), reactions needed for the activities portrayed in the picture can take place at reasonable rates. The reactions going on inside the picnickers need a slightly higher temperature, of around 37 °C, if they are to function; nevertheless, these temperatures are all much the same. Their closeness is even more striking if we express them on the absolute temperature scale, which we have already seen is the fundamental scale to use in thermodynamics. On this scale 20 °C corresponds to 293 K and 37 °C corresponds to 310 K; so both temperatures are within a few percent of 300 K. Henceforth we shall deal with temperatures on the Kelvin scale, and take "normal temperature" to be 300 K.

The things going on at the picnic depend on chemical reactions, both for the physical acts of ingestion and for the states of the brain that correspond to enjoyment. In order for any reaction to proceed, enough energy has to assemble at the molecular site where the atoms are undergoing reorganization. The higher the temperature, the more likely it is that random accumulations of energy will be large enough to allow the atoms of the relevant molecules to explore new arrangements. If the temperature becomes too low, the processes of life will cease, for now atoms are literally frozen into their present arrangements. Then they cannot explore, just as water cannot flow when cooled below its freezing point. This cold world (below) is the one we shall explore first.

At low temperatures (not far below 300 K), most chemical activity ceases, because atoms are frozen permanently to their neighbors and cannot accumulate enough energy to explore alternative partners.

Catching Cold

How can we achieve cold? How can we cool, for instance, the components of the picnic to one-tenth their normal temperature? How can we cool them to 30 K (to −243 °C), when cooling seems to be against Nature?

Of course, cooling is *not* against Nature as expressed by the Second Law. That forbids (or, to be fair, declares exceedingly unlikely) the *spontaneous* shift of heat from cold to hot without there being change elsewhere. It does not forbid the transfer of heat against a temperature gradient (from cold to hot, as from the inside of a refrigerator to the room outside) if change occurs elsewhere. There is another analogy (which we shall develop further in Chapter 8). Just as cold cannot emerge spontaneously, so the emergence of the ordered forms characteristic of life cannot arise spontaneously: but the Second Law is not contravened by the emergence of life if compensating change takes place elsewhere. In order to live, we have to eat, which means we have to destroy the ordered, high-quality form of energy imported in, for instance, a sandwich. In order to cool, we have to destroy the quality of energy elsewhere, such as by burning a lump of coal, bursting a nucleus, or letting water gush through the tubes and turbines embedded in dams. That destruction of the quality of the world's energy may be geared to drive a local abatement of chaos, and just as a cathedral may be built by doing work, so an object may be cooled.

The easiest way to see how to go against Nature locally but not in the large is by means of one of the engines we have considered already. However, now we run it backward by coupling it to a slightly more powerful engine (see figure on left). We have seen that work may be produced if heat passes from hot to cold; we shall now see that work may be used to transfer energy as heat in the opposite direction. We shall use the Carnot cycle to make the point, but other cycles are used in practical, work-producing engines, including those in refrigerators.

Point *D* in the indicator diagram (on the facing page) now represents the starting point of the cycle. The piston is pushed out by the gas trapped inside the cylinder and does work on the outside world. As it does so, its temperature tends to fall, because the energy stored in the motion of the gas particles is being converted into work. However, since the cylinder's walls are conducting at this stage of the cycle, the process is isothermal, because energy jostles in from the object being cooled (see figure on page 132). This stage takes the system to *C*.

At *C* the crank has rotated so far that now the piston begins to compress the gas. This takes place while there is no thermal contact with the cold reservoir (the object being cooled), and so the temperature of the gas rises. With rising temperature the pressure rises too.

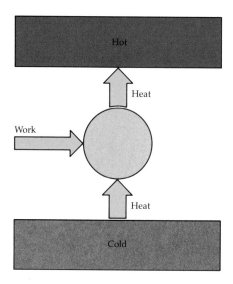

A refrigerator is an engine that is driven in reverse. So long as work is done, heat may be transported from a cold source to a hot sink.

The indicator diagram for a Carnot engine run as a refrigerator. This is the same as in the figure at the bottom of page 18, but the cycle is traversed in an anticlockwise direction.

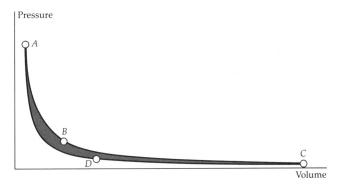

At *B* thermal contact with the hot sink is established. The high-temperature reservoir is now the *sink*, because that is where the refrigerator is dumping the energy it is withdrawing from the object being cooled. The dumping stage is the reason for all the coils at the back of a typical domestic refrigerator: they make contact with the surrounding air, which constitutes the hot sink. The crank continues to turn, for it is being driven by another engine. As it turns the piston is driven into the cylinder. This is an isothermal compression stage: the gas does not become hotter, because the energy of the gas particles jostles out into the hot reservoir (see figure on next page). This is the source of the warmth of the coils. Because the temperature of the confined gas is higher than in the expansion phase, *more* work has to be done to compress the gas than was generated as it expanded.

The work-consuming isothermal compression takes the gas to *A*. The thermal contact is then broken, and the turning of the crank now enters the final, adiabatic, expansion stage of the cycle. In this step from *A* to *D*, the gas drives out the piston, doing work on the outside world. In the process the gas cools, because its particles are surrendering their kinetic energy in their collisions with the face of the receding piston. The temperature of the gas in the cylinder drops to that of the object being cooled, and the device is back at its initial state, ready to be driven around the cycle again. Note that it has to be *driven* round the cycle, because the work involved in compressing the high-temperature gas, and almost literally squeezing its stored energy out into the hot reservoir, is greater than the work coming from the isothermal expansion step. We have transported energy from the cold to the hot; but we have had to do work to bring the transfer about.

In order to cool something, to go against Nature locally, we not only have to organize equipment to *do* work (we need a motor), but also have to

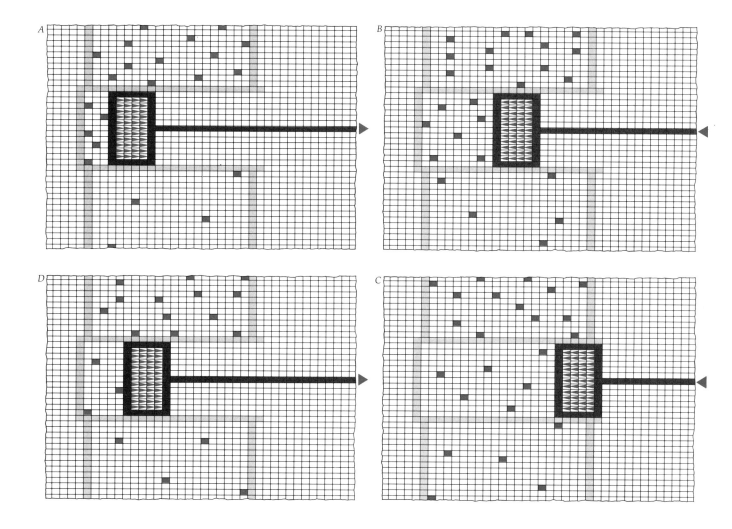

arrange for it to be coupled to an engine on which work can be *done* (the motor drives a compressor). This is the deep reason why refrigerators were a relatively late addition to our collection of kitchen equipment: heating is natural; cooling has to be organized.

But *how much* effort is needed to extract a specified quantity of energy as heat, and then keep the object cold? This is a very important question, and the answer to it has implications for the ease of our journey through powers of temperature. The Second Law allows us to arrive at a numerical answer. The argument runs as follows.

The Mark II model of Carnot refrigeration. In the stage from D to C, the ON atoms drive out the piston, but the proportion ON remains constant, because energy can jostle in from the cold source. The stage from C to B is an adiabatic compression, and atoms are turned ON as they are stimulated by the incoming piston. The stage from B to A is an isothermal compression: although more atoms are turned ON, the ON-ness jostles out into the hot sink. This step in effect squeezes the energy picked up from the cold source out into the hot sink (and work is required). The cycle is completed by the adiabatic expansion from A to D, which lowers the temperature of the gas to that of the cold source.

Any natural process must generate at least a tiny amount of entropy in the Universe. But when heat is absorbed from the object being cooled, there is a *reduction* of its entropy. The magnitude of the reduction is given by the expression that we first encountered in Chapter 2, namely, *(Heat absorbed)/Temperature*; here *Temperature* means the temperature of the cold object, $Temperature_{COLD}$. Entropy has been reduced because thermal motion has been reduced in the cold source. Now, in order for the entropy to increase overall, some entropy must be generated elsewhere. That is the thermodynamic reason for dumping heat into the room, because the supply of heat corresponds to a generation of entropy. This positive contribution to the overall entropy is *(Heat dumped)/Temperature*; where *Temperature* now means the temperature of the hot sink, $Temperature_{HOT}$.

We are now at the crux of the argument. Since $Temperature_{HOT}$ is greater than $Temperature_{COLD}$, the entropy generated is greater than that destroyed only if *Heat dumped* is greater than *Heat absorbed* (see figure below). We can dump more heat only if we can somehow add to the flow of energy as it goes from the cold object to the warm room. That can be achieved by contributing energy by doing work, and that is why we need to do work in order to bring about refrigeration. The *minimum* work we must do will augment the stream of energy just enough to generate slightly more entropy than was destroyed. The two entropy values are equal (see figure below) when the value of *(Heat absorbed)/$Temperature_{COLD}$* is equal to the value of *(Heat dumped)/$Temperature_{HOT}$*, where *Heat dumped* is the energy drawn from the cold object plus the energy supplied as work, that is, *(Heat absorbed) + (Work done)*. By equating the two expressions, we can ar-

The illustration on the left shows that the entropy of the cold source decreases and that of the hot sink increases during refrigeration. We can ensure that overall there is an increase if we add to the stream of energy by contributing some as work. The augmented stream is shown on the right, and even though the temperature at which it is stored in the hot sink is higher, the increased quantity ensures that the entropy has increased overall.

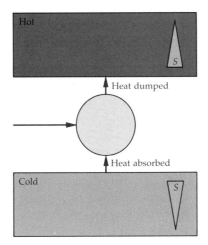

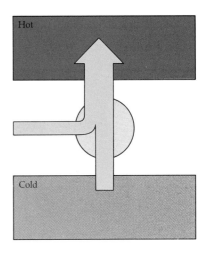

rive very easily at the following expression for the minimum amount of work we must do to extract a given quantity of energy as heat:

Minimum work = (Heat absorbed) × {(Temperature difference)/Temperature$_{COLD}$*}.*   (7.1)

By *Temperature difference* we mean *Temperature*$_{HOT}$ − *Temperature*$_{COLD}$. As always in thermodynamic arguments, the qualification "minimum" means the value after we allow for all other losses, such as may arise from leaks, friction, or faulty design. These all worsen the requirements: the expression gives the *absolute* minimum, in a world of perfect materials and ideal (and quasistatic) processes. In the real world, somewhat more work is needed to extract any specified quantity of energy. We shall refer to the expression in curly brackets as the *Carnot factor*.

We have calculated the work needed for *producing* cold. Now we calculate the power input we need in order to *sustain* that cold. A cold object cannot be completely isolated; so there is always a leakage of heat into it from its warmer surroundings. The *rate* at which energy leaks in as heat is proportional to the temperature difference, and the constant of proportionality depends on the quality of the insulation and the size of the object. The rate at which we do work must therefore be sufficient to counter this flow of heat (see figure on left), and is therefore proportional to the Carnot factor. Overall, therefore, the minimum rate of doing work, the minimum power we must expend is

$$Power \propto (Temperature\ difference)^2/Temperature_{COLD} \qquad (7.2)$$

This expression has some implications for the power required to keep an object's temperature close to absolute zero. We are dividing by *Temperature*$_{COLD}$; so as it gets closer to zero, the minimum power required rises toward infinity. The whole output of the world's generating plants could not keep something at absolute zero. Furthermore, even the work involved in getting that cold is infinite, because the Carnot factor is close to infinity when *Temperature*$_{COLD}$ is close to zero. (This is another aspect of the observation that a Carnot engine has to become infinitely big if one of its reservoirs is to operate at absolute zero; see *Carnot* in Appendix 3.)

An initially more encouraging (but soon to be discouraging) aspect of the expression concerns the amount of work that has to be done to cool an ordinary object under ordinary conditions. Suppose, for instance, that we wanted to remove 1,000 joules of heat from a tray of water in order to make ice, and that the temperature has already fallen to the freezing point (0 °C, 273 K). Suppose the refrigerator is standing in a room at 20 °C (293 K); then the temperature difference is 20 K, and the amount of work we must do is only 20/273, or 0.073, times the amount of energy we want to remove. Removing 1,000 joules of energy therefore demands doing only 73 joules of

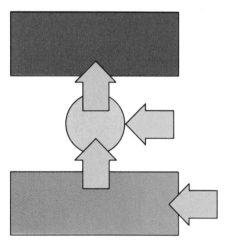

The problem with sustaining *a low temperature is the leakage into the cold source of heat from outside. This flow must be balanced by the refrigerator, so if heat leaks in at a particular rate, work must be expended (as determined by the Carnot factor) fast enough to remove it again.*

work. When the refrigerator is running, 1,073 joules of energy is dumped into the room at the expense of 73 joules of work (consumed in the form of electricity), the balance of 1,000 joules coming from the water we are freezing.

This calculation is encouraging; it stimulates the following thought. For the modest investment of a few joules of energy, we are getting a large quantity of heat (73 joules was invested; 1,073 was obtained). But very little energy is stored in a pint of water inside a refrigerator. Suppose we think on a larger scale, and consider the world outside: the backyard, a river, or a lake. Using the outside as a heat source, we could invest 73 joules of energy and achieve a healthy 1,073 joules of heat out of the back of the refrigerator virtually forever. This is the basis of the *heat pump*, which is nothing more than a large refrigerator, but now interest is focused on what is coming out of the coils at the back, not on what is being cooled (which is now the outdoors).

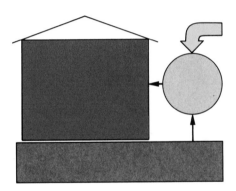

A heat pump *is a refrigerator in which interest is focused on the coils at the back. Work (yellow) is supplied to pump heat from the cool surroundings into the warm house. A heat pump focuses energy.*

Apart from the expense of its initial installation, the heat pump is an attractive option for the householder. It is in effect an energy *magnifier*; by investing a little energy as work, we gain a large amount of energy as heat. In the example we are considering, where the heat is extracted from the cold (0 °C) ground and supplied at a modest temperature (20 °C) indoors, the ratio of heat output to work supplied is 1,073/73 = 14.7, or 1,470 percent conversion! Furthermore, there is a moral dimension to the use of heat pumps. It is ecologically and economically senseless to take the high-quality energy stored in fossil fuels and to use it merely as heat, the lowest of its possible uses (the lowness depending on the temperature at which it is frittered away). It is far better to use the minimum amount of this highly concentrated, five-star energy to gather together some of the old, low-quality, ragged energy lying around in our own backyards, and to focus it into the house. This is efficient energy management!

As always, there is another side to the ecological aspect. At present, with the widespread use of fossil fuels, we are living off our inheritance from the past. The energy we are using, for whatever purpose, is the accumulated total of ages, and its removal from a locality has only indirectly damaging effects. But if the use of heat pumps were to become widespread, we would be using the energy of the present. In particular, we would be using the energy provided that day by the Sun, energy that other processes of life may also need. We would be cooling the ground around us for longer periods of the year, and lowering its average temperature. No one really knows whether the capturing of warmth on this large scale from the immediate and present environment would have long-term ecological effects, such as retarding the germination of seeds and the breeding of earthworms.

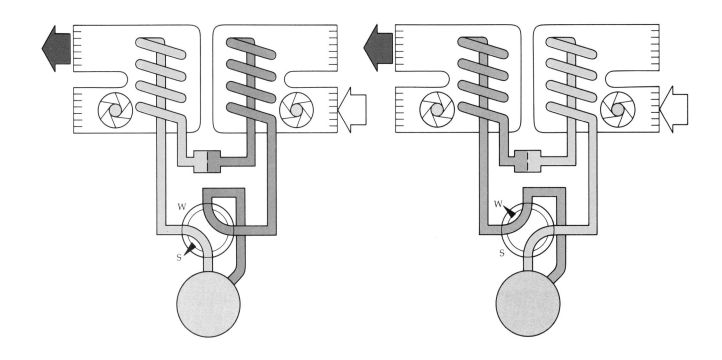

A heat pump may be used as an air-conditioner during summer. The exterior is on the right of the apparatus; green denotes high-pressure gas, and yellow denotes low-pressure gas; cooling occurs by expansion at the nozzle. On the left the equipment is in the summer configuration, and the flow of working fluid warms the exterior (the hot sink) and cools the interior (the cold source). These roles are reversed in the winter configuration on the right.

That doubt aside, another advantage of heat pumps is that the same one used as a heater during the winter can be run in reverse and used as an air-conditioner in the summer! Simply by turning a valve on an arrangement like that shown above, we can reverse the flow of the working fluid, so that the interior of the house becomes the cold source, and the exterior the hot sink.

Ordinary household refrigerators operate in the vicinity of room temperature. If we compare the temperatures of ice (273 K at its melting point) and a typical room (298 K), we see that they are almost the same, and certainly well within a power of ten of each other. Ice and balmy air, and human bodies too (whether alive or dead), are in the same thermal region of physics; in it energy can accumulate by chance to allow many chemical reactions to proceed. In order to go the full power of ten downward in temperature, we have to go from 300 K to 30 K, which is below the boiling point of air, below even the average surface temperature of Pluto, which is estimated at between 40 and 50 K. That is, we are considering temperatures that seem not to appear naturally in the matter of the solar system. In order to achieve them, we have some serious work to do, and some sophisticated machinery to devise.

The First Power Down

Some idea of the work needed to reach a temperature of 30 K can be obtained by considering the implication of expression 7.1 (on page 134). Suppose we decided to extract 1,000 joules of energy from an object at 30 K and to dump it directly into the warm world at 300 K. This represents a large increase in the quality of those 1,000 joules, because now the energy is stored at a high temperature, and to create such an increase in quality, we must destroy at least as much quality elsewhere. In other words, a great deal of work has to be generated and transferred to the refrigerator. The expression (7.1) tells us how much: the ratio of temperatures is 300/30 = 10; the Carnot factor is therefore 9; so the work we have to do is 9,000 joules. Furthermore, maintaining a cold object at 30 K in an environment at 300 K requires 910 times as much power as when the cold object is to be maintained at 0 °C: a 100-watt refrigerator would have to be upgraded to 91 kilowatts! (It is cheaper, of course, to improve the insulation and to use a tiny sample.)

The contriving of the cooling also demands a more intricate strategy than is required near the temperature of the original picnic. There are often several steps involved.

One basic technique is to adapt a step of the Carnot cycle. There we saw that the adiabatic expansion of a gas reduced its temperature, because the particles striking the surface of the piston gave up their energy. This energy is transmitted to the outside world as work, and as it is given up, the particles' thermal motion becomes less vigorous. The temperature of the gas therefore falls. However, there are various difficulties with cooling gases in this way. First, only a small temperature drop results from a reasonable change of pressure, and the drop gets smaller as the operating temperature gets lowers. Second, since the process depends on a machine built from moving parts (such as a piston sliding in a cylinder), there are problems arising from lubrication and noise. Nevertheless, the technique is used as an initial stage in some commercial devices, and it produces cool gases for use in the stage we are about to describe.

An effective method of cooling a gas, provided it is already fairly cool (hence the need for an initial cooling stage), is to rely on an effect that is analogous to sending interplanetary missions off into outer space. The link can be explained as follows.

In order to send a space probe away from Earth, we have to give it a high initial kinetic energy by an initial burn of its rockets. (A baseball player hitting a ball does the same thing: his body briefly burns, and the ball receives a high initial kinetic energy when it is struck.) After the burn the rockets are extinguished, and the spacecraft is carried upward by its

When a gas expands, there is a net separation of the particles, which is the analog of a fleet of spacecraft climbing away from a collection of Earths. Because the particles attract one another, they slow down as they separate, which corresponds to a lowering of the temperature of the sample.

own momentum. As it climbs it gains potential energy, because it comes increasingly far from the center of the Earth. Since its energy is constant (the engines no longer burn), its kinetic energy must decrease as it is converted into potential energy; so the craft must slow down as it climbs. Therefore, in order to escape the gravitational field of the Earth, the craft must be given at least a certain minimum kinetic energy in the first place. This *escape velocity* is the speed an object must be given to escape from the Earth; it is 11.3 km/sec, or 25,300 mph.

Gas particles are like rockets and Earths in the following sense (see figure on left). Two particles near each other experience an attractive force that tends to hold them together, as gravity holds the rocket and the Earth together (but now the force is electromagnetic, not gravitational). Therefore, if one particle is to move away from the other, it has to climb out of the other's attractive force field. Each particle is the spacecraft to the other particle's Earth. The particles may be moving quickly immediately after they have collided (as in the figure on the left), but just like the spacecraft, they lose speed as they separate, because their potential energy is increasing at the expense of their kinetic energy, and they have no rockets to burn.

The climbing apart of particles, and their consequent slowing, is occurring throughout a gas. But there are also many particles moving *toward* each other. These are like spacecraft plunging to Earth, and speeding up as they fall, because now potential energy is being turned into kinetic energy as the particles move together and experience each other's attraction. On the average, there are as many particles falling together as there are particles climbing apart; so the *average* speed of the particles taken overall remains constant. Average speed is related to temperature (p. 56), and so the temperature of the gas remains constant. But suppose the gas is not in a container of constant volume, and is expanding. Then the quest for chaos ensures that the particles move into whatever space is available. Therefore all the spaceprobes and Earths wander off into the increased volume that is becoming available, but in doing so they have to climb away from each other. Now, on the average, more are climbing apart than are falling together. Consequently, on average, they slow. Since lowering of average speed implies cooling, the gas cools. This is the *Joule-Thomson effect* (Thomson being, recall, Kelvin's other name).

It should be noted that the Joule-Thomson cooling effect is quite different from the cooling brought about by adiabatic expansion. In adiabatic expansion *work* is done, and even if there are no attractive forces between the particles, energy is lost to the outside world. Here work need not be done, but the gas, so long as there are attractive forces between its particles, will cool. The tendency to chaos implies the increasing separation of

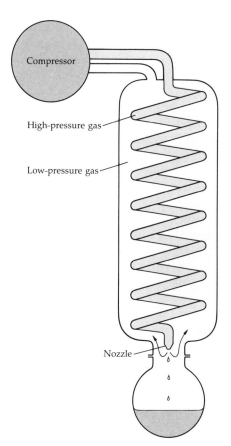

A schematic diagram of the Linde refrigeration process *for liquefying gases. The gas is compressed and undergoes Joule-Thomson expansion at the nozzle. The cooled gas is recirculated, and with successive expansions its temperature is lowered further. In due course the temperature falls to its boiling point, and then liquid begins to drip out of the nozzle.*

the particles, and in their stumbling into greater freedom some of their kinetic energy is converted into potential energy. As that kinetic energy decreases so too does the sample's temperature. Chaos this time has generated cold.

The Joule-Thomson effect is used commercially to step down through the first power of ten from normal temperatures. Because each expansion of the fleet of particles reduces the gas's temperature only a few degrees, the gas is recirculated in an apparatus like the one shown on the left, and successive steps down the staircase of temperature finally bring it to below its boiling point. Then liquified gas drops out of the nozzle.

Once we have liquid air, which under normal pressures forms at about 80 K, we are in the realm of the first power of ten of cold. We can approach the center of the realm by using this cold liquid air to cool hydrogen, and then using that cold hydrogen in a Joule-Thomson apparatus. Liquid hydrogen drips out of the nozzle when successive cycles of cooling have reduced its temperature to 20 K.

Now we are in a cold, dead world, where all the chemical (and therefore biological) activities of the picnic are dead. The atoms in the molecules of the materials still wobble, but they have too little energy ever to change their partners. All chemical change has been eliminated: molecules, and the materials they constitute, remain forever in suspended inanimation.

The tendency to disperse has not been eliminated, but the opportunities to disperse have been blocked. Now there are only small, feeble, fleeting fluctuations of energy accumulating in bonds, and so the atoms are effectively frozen into their locations. This cold, frozen world, however, is still in the world of ordinary physics: although the opportunities for molecular rearrangement have been closed, the solids still ring with noise. Although the atoms are trapped, they are not still, and vibrations still rattle through the latticework they form. Chemistry is dead, but for physics the activity of the lattice is still qualitatively like that of our own thermally turbulent world. In order to reach a new domain of physics, the subtle, quiet, physics normally masked by the brash motion of atoms of everyday physics, we have to lower ourselves through another power of ten.

The Second Power Down

In order to remove another 1,000 joules of energy and step down from 30 K to 3 K, we have to do another 9,000 joules of work. Notice how the step in degrees is decreasing even though we do the same amount of work: the first 9,000 joules took us down through 270 degrees; the second takes us only through 27. In order to maintain a sample of 3 K in a world at 300 K, and in something like a domestic refrigerator, we would have to expend nearly 11,000 times as much power: 100 watts becomes over a megawatt. Of course, we take extreme precautions, use more insulation, and deal with smaller samples.

We take the step from 30 to 3 by using helium. First we cool warm, everyday helium (which bubbles out of the ground in Texas along with natural gas). We can cool helium either by putting it in contact with liquid nitrogen or liquid air, or by making it do work and thus cooling it by adiabatic expansion. Then the cool helium is passed repeatedly through a Joule-Thomson device, and in due course the liquid drips out of the nozzle. The boiling point of helium is 4.2 K, and so liquid helium is two powers of ten down from the picnic.

Now we are in the world of thermal quiet, and the networks of atoms in solids no longer ring with thermal noise. Here it is quiet enough to hear a new physics in action, a physics that at higher temperature is overpowered by the turmoil of activity of the atoms. Now it is quiet enough for *superconductivity*, the property that some substances have of conducting electric currents without resistance.

Superconductivity was discovered by the Dutchman Kamerlingh Onnes in 1911. Its significance was summarized by M. Zemansky as follows.

> Of all the peculiar things that happen at low temperatures, superconductivity is:
>
> 1. The most *spectacular* (persistent electric currents in metal rings should last over 100,000 years);
>
> 2. The most *useful* to physicists and engineers (it is possible to make superconducting magnets, thermal switches, frictionless gyros, and small, fast, zero-power-consuming computers); and
>
> 3. The most *challenging* to theoretical physicists (superconductivity was unexplained for 46 years until Bardeen, Cooper, and Schreiffer in 1957 made the first theoretical breakthrough).

One technological application of superconductivity is to generate the magnetic fields required to confine plasmas in controlled nuclear-fusion devices. Sustaining a low temperature requires power, which is a drain on the overall output of the reactor. Good design, however, can reduce the power required to around 10 megawatts in a reactor producing a million megawatts of power.

At only a little below 4 K, at 2.2 K, a sister property of superconductivity, *superfluidity*, makes its appearance. Just as in superconductivity electron flow persists indefinitely, so in superfluidity (a property shown only by helium) the flow of the atoms themselves continues for ever. This is the world of flow without viscosity, and where the thermally quiet liquid creeps around apparatus and through capillaries, exploring everywhere.

It is very tempting to linger with these strange but useful properties, but we can only mention them (the bibliography list some works in which these topics can be explored further). Before we move still lower in temperature, we need to mention that 3 K is a temperature with a much broader significance: *3 K is the temperature of space*. Our picnickers in the top figure on page 128 are brushed by radiation. If we discount the intense radiation from the Sun, the unseen radiation generated by cosmic rays, the natural radioactivity of the Earth, and the long-wavelength radio and TV emissions that bathe us all, there remains a delicate wind of radiation of wavelength about 3 cm. This is the radiation first detected in the careful experiments of Penzias and Wilson, when they listened under conditions of great quietness to the messages from space.

The radiation detected by Penzias and Wilson is the cosmic *microwave background*. It covers a range of wavelengths, but peaks close to 3 cm, and has the characteristics of the radiation that would be emitted by an object at a temperature of about 3 K (more precisely, at 2.7 K). This is what has become of the Big Bang. In current cosmological models, the radiation and

The microwave antenna used by Penzias and Wilson to detect the cosmic microwave background radiation. The antenna is sited at the Bell laboratories in New Jersey. The detector itself has to be kept very cold, so that the electrical noise it generates does not obscure the weak signal.

matter of the early Universe severed their mutual thermal contact when the Universe was around 700,000 years old, when its temperature had dropped to about 3,000 K. Ever since, the Universe has gone on expanding, and the waves of radiation in it have been stretched. This stretching has increased the wavelengths of all the radiation present, and they are now so long that most of them are around 3 cm. But the matter of the Universe, since it is no longer in thermal contact with the radiation, has not cooled as much. This is why the scene at the picnic is really two worlds. There is the world of warm matter (which in that locality happens to be at around 300 K, but a few light-minutes away there is a much hotter globule of matter, the Sun, that provides the warmth). There is also the invisible world of the cosmic background, the cold, all-pervasive world of 3 K radiation. All the scenes of life are played out against this cold background: we actors are hot specks on the cold stage of space.

Lower Powers Down

Now we step down another power of ten, from 3 to 0.3. In order to get to such depths of cold we use two steps: the first takes us to 1 K, the second from there to our goal.

In the first step we cool in the same way that we ourselves cool on stepping out of a swimming pool. The water molecules clinging to our skin drift off in chaos; but, in order for them to be free, the bonds that cause them to adhere to us and to each other must first be broken (see figure below). The energy required can wander in from our body; once it is in, the liberated molecule can fly off into freedom, with very little chance of ever returning its stolen energy. Hence we cool. Exactly this process was used in the earliest refrigerators: toward the middle of the eighteenth century William Cullen, professor of chemistry at the University of Edinburgh,

Liquids can be used to cool by evaporation. In the illustration on the left, there is a thin film of liquid on a warm surface (which could be the bulk of the liquid). Particles need to overcome their attractive interactions with the surface and with their neighbors, and so need energy to escape. Those that do escape carry energy away with them, and so the atoms of the surface turn OFF.

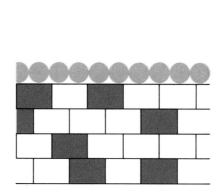
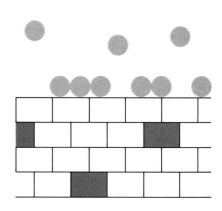

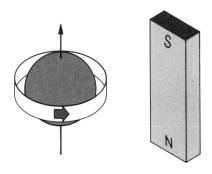

An electron possesses the property "spin." For our purposes we can imagine it as a classical spinning motion, but each electron must be thought of as spinning at exactly the same rate. Since electrons are charged, and moving charges generate magnetic fields, the spin of an electron causes it to behave like a tiny bar magnet, as in the illustration on the right.

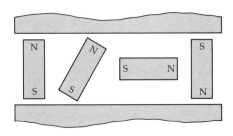

According to classical physics, an electron's magnet could take up any orientation in a magnetic field (and different orientations differ in energy): four are shown here. According to quantum theory, however, all but two of these orientations are forbidden. Only the two outermost in the illustration are allowed.

invented a refrigerator based on the cooling produced by pumping away water vapor. As water cools when it evaporates, so two powers down does liquid helium. The atoms at the surface of a sample acquire enough energy to break free of the feeble bonds that hold them to their neighbors. As a result, they escape with their energy to become a gas. The liquid left behind, now possessing less energy, is cooler than before. This process of vaporizing at least some of the hard-won liquid helium is a sacrifice that lowers the temperature to the intermediate stage of 1 K. Now 0.3 K moves into range.

The second step of this sequence involves a clever new deployment of the Second Law. In order to see what is involved, we need to look a little into *magnetism*, for this is the handle that the technique grips.

Magnetism is due to the *spin* of electrons. (Although "spin" is a subtle concept, we can for now think of it as an ordinary spinning motion.) Because each electron is charged, and moving charges give rise to magnetic fields, the spin of the electron gives rise to a magnetic field. Each electron therefore behaves like a tiny bar magnet (see top figure on left). The special property of an electron, however, and one that distinguishes it from an ordinary bar magnet (which consists of billions of spinning electrons), is that in a magnetic field it may take up only *two* orientations (see bottom figure on left). (This is a consequence of quantum theory.) The two orientations are called spin "up" and spin "down"; we will call them UP and DOWN. In most materials there are exactly equal numbers of UP electrons and DOWN electrons, and we say that in these compounds all the electrons are *paired*. Since the magnetic fields of paired electrons cancel, it follows that most materials do not give rise to a net magnetic field. However, some substances have unequal numbers of UP and DOWN electrons. These have net magnetic properties, and are called *paramagnetic* materials.

A simple model of a paramagnetic material is illustrated in the top figure on the next page: it consists of a lattice, in each cell of which is an electron (which may be UP or DOWN) outside a core composed of paired electrons. In practice a compound such as gadolinium sulfate is used in the manner we are about to describe. In gadolinium sulfate each gadolinium ion carries seven electrons, which spin in the same direction, and each ion is separated from its neighbors by a sheath of sulfate ions and water molecules. This ensures that each set of seven electrons behaves as a single, almost isolated unit. The model compound in the top figure on the next page is much simpler, because there is only one unpaired electron at each site, and thus captures the essential features without getting bogged down in the details of the actual compound.

The link between this discussion and the Second Law should be apparent from this figure. We see that the presence of the electron spin allows another way to achieve chaos: some spins may be UP; others may be

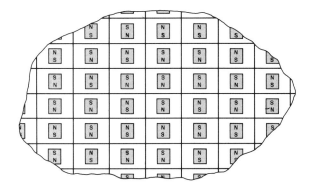

A paramagnetic material consists of particles that carry unpaired electrons. In the absence of an applied magnetic field, there are equal numbers of spins UP and DOWN.

DOWN. The Demon has more freedom of action, for it may now switch spins UP and DOWN as well as ON and OFF. This new contribution to chaos counts as a new contribution to entropy. Moreover, with an adroit deployment of the Second Law, we can use the magnetic chaos to retreat into low temperatures.

Suppose we take a sample of the compound. In it, at some arbitrary temperature, there will be equal numbers of UP and DOWN electrons. The particles will also be vibrating with thermal motion; so some will be energetically ON, others OFF. In the figure below, let us mark UP-ness as yellow, DOWN-ness as green, ON-ness as red, OFF-ness as white. At this stage the sample has an entropy that arises from both the many ways to arrange ON-ness and OFF-ness and the many ways to arrange UP-ness and DOWN-ness.

Now suppose that a field is applied to the sample by an external magnet. Different orientations of a bar magnet in an applied field correspond to different *energies*; so when that field is present UP-ness and DOWN-ness also correspond to different energies, as well as to different orientations of

The Mark II universe model of a paramagnetic material at some arbitrary temperature. In the absence of a magnetic field, the opportunity for electrons to be UP (yellow) or DOWN (green) is an additional contribution to the entropy of the sample. Atoms may still be ON (red) and OFF (white), but the spins they carry may be UP or DOWN at random. The two illustrations show the same distributions of ON and OFF, but different distributions of UP and DOWN.

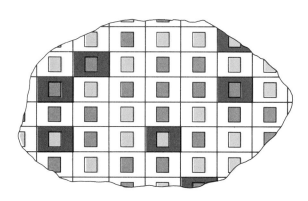

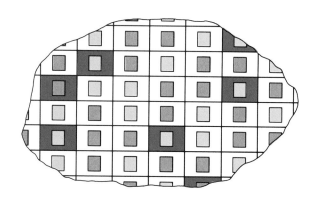

the electron spins. For simplicity, let us for now disregard the ON/OFF contribution to the energy, and concentrate on the UP/DOWN contribution. With a magnetic field present, UP corresponds to higher energy than DOWN.

If the ratio of the number of UP electrons to the number of DOWN electrons were to remain the same as the field is increased, the temperature would be reported as rising. In fact, we would have to report the system as having an infinitely high temperature! Why? Because the expression of temperature is:

$$Temperature = 1/\log (Number_{DOWN}/Number_{UP}).$$

Since $Number_{UP}$ is equal to $Number_{DOWN}$,

$$Temperature = 1/\log (1) = 1/0 = \infty;$$

that is, the temperature at this stage is infinite. It is a general result, which we shall explore later, that if two states differ in energy but are equally populated, then the temperature of the system is infinite.

Since the electrons are now infinitely hot, energy jostles out of them (see figure below). Energy jostling out corresponds to UP particles changing to DOWN. The tendency to chaos is claiming its own again, as energy jostles out. But note the secondary consequence of this process. Spins are flipping from UP to DOWN; so when thermal equilibrium has been established with the surroundings, there are more DOWN particles than UP. This means, in turn, that the magnetic fields arising from the electron spins no longer cancel, and the sample now gives rise to a net magnetic field. In other words, the sample is now *magnetized*. This process, which is gov-

The model of the isothermal magnetization stage. Initially (on the left) the numbers of atoms UP and DOWN are the same. A magnetic field is applied while the sample is in contact with a thermal reservoir. The energy difference that now arises between UP and DOWN (with DOWN lower) means that energy jostling out corresponds to spins switching DOWN. At the end of this stage, on the right, there are more DOWN spins than UP, and energy has wandered off into the reservoir.

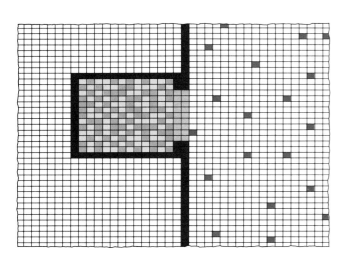
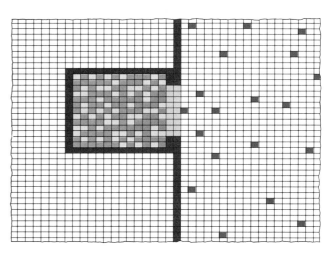

erned solely by the usual tendency of energy to disperse, is called *isothermal magnetization*.

Having magnetized our sample, we now break the thermal contact with the surroundings; so the next step is adiabatic. No energy can now escape or enter as heat. Consider what happens, in these circumstances, as the magnetic field is slowly and carefully (quasistatically) reduced to zero. When the field reaches zero, we can expect the magnetization of the sample to have dwindled to zero too, because now there is no favored direction for the electron spins. Since the process of losing magnetization is adiabatic, this step is one of *adiabatic demagnetization*.

The adiabatic demagnetization step cools the sample in the following way. First, the step occurs at constant entropy. That must be so because it is done quasistatically (so there is no turbulence, no generation of entropy); yet no heat enters or leaves the system, because we have arranged for it to be adiabatic by removing all thermal contact. When the magnetic field is absent, the spins can be UP or DOWN willy-nilly, because their energy is independent of their orientation. In contrast, when the field was present, a certain proportion had to be UP because the temperature was definite, and the different orientations of the spins corresponded to different energies. Therefore, when there is zero magnetic field, the Demon can deploy more than when the field is present (see figure below). Hence, *for the spins*, the entropy has increased during the demagnetization step: the Demon discovers more freedom in the absence of the field, because then UP-ness and DOWN-ness are dissociated from considerations of energy. But *overall* in the sample the entropy is constant, as we have already stressed; so the

The model of the adiabatic demagnetization step. Initially the spin distribution is the one achieved in the magnetization stage. The thermal contact with the reservoir is broken, and then the magnetic field is reduced to zero. In zero magnetic field, UP and DOWN spin states are equally populated (righthand illustration). The spin entropy has increased, but overall, since the adiabatic demagnetization is quasistatic, the total entropy must remain constant.

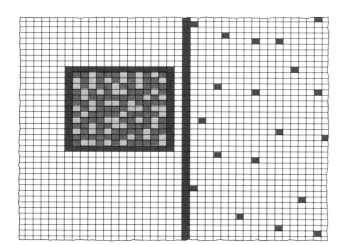
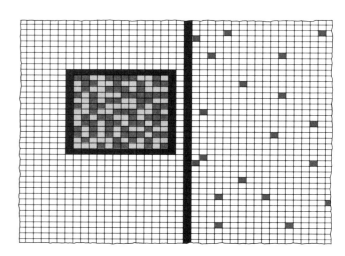

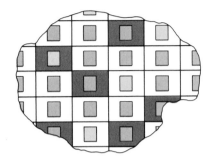

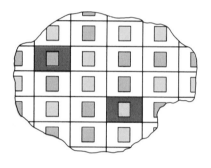

The constancy of the entropy is achieved by atoms turning OFF as spins turn UP. The initial state is on the top, the equal-entropy but thermally extinguished state on the bottom. This step has therefore cooled the sample.

increase in entropy for the spins must be offset by a reduction of entropy somewhere. The only place where that could occur is in the thermal motion of the atoms, the background ON-ness and OFF-ness that we have been ignoring so far. That is, in order for the same overall entropy to be maintained in the sample, the entropy of the thermal motion of the atoms must decrease; so the thermal motion itself must decrease, and that decrease lowers the temperature of the sample (see figure on left). It follows that an external observer will report that the system has cooled. The electron spins have acted like a tiny refrigerator embedded in the sample, and under the influence of the magnetic field have pumped energy from the atoms out into the surroundings.

By means of the technique of adiabatic demagnetization, solids have been cooled not only down to 0.3 K, but even down two powers of ten still deeper into coldness, to temperatures around 0.003 K. Now even physics is beginning to stop in the almost perfect quiet of the interiors of solids, where the movement of atoms is no more than an occasional rustle. But this is not the end of the trail downward: we now know how to silence motion even further. Instead of using electrons as refrigerators, we can use the spins of the nuclei of atoms. Many types of nuclei spin like electrons, and, being charged, behave like tiny bar magnets. They too may be used to cool. By means of this technique of *nuclear adiabatic demagnetization*, we have reached the lowest temperature certainly in the solar system, possibly in the Galaxy, and even perhaps in the Universe, for such almost infinitely Arctic cold is a sign of very advanced civilization, since the natural direction of change has to be overcome so completely. The world (and perhaps the Universal) record in 1994 stands at 2.8×10^{-10} K, twelve powers of ten down from the picnic. In this almost perfect quiet, all physics is almost perfectly dead.

Powers Hotter

Now is the time to heat the people at the picnic, and to explore what happens as we raise the temperature tenfold in each step. What happens when we throw the switch that takes the picnickers from 300 K to 3,000 K?

They burn. This temperature is high enough to melt all but a handful of materials (tungsten, the metal with the highest melting point, melts at 3,387 K, corresponding to 3,114 °C). Now so much energy is available that atoms linked to each other by bonds hardly depend on chance accumulations of energy to gain freedom to explore, but can explore almost at will. In particular, the atoms of those delicate, complex molecules that constitute living things are liberated, and the molecules collapse into simple forms. Where there were proteins, there is now carbon dioxide; and even the carbon dioxide is disrupted and broken into atoms. Chaos has found almost full chemical liberation: atoms can stumble anywhere; molecules are entities of the colder past.

A temperature of 3,000 K, where chemistry has been burnt away, is also significant in being the temperature at which many electrons become liberated from their parent atoms, because now even they have enough energy to explore regions distant from their nuclei. The picnic party is now a *plasma*, a gas of nuclei and electrons. Moreover, because the charged particles interact with electromagnetic radiation, and emit and absorb light, the world becomes opaque. Because the hot charged particles can jostle both each other and the photons in the electromagnetic field, energy wanders freely between matter and radiation. No longer is the electromagnetic field at a different temperature from the matter: now their temperatures are the same. Thermal contact between the two has been established, and the Universe is a thermal unity again.

I say *again:* the Universe is globally at thermal equilibrium (at least locally, where we are heating) *again.* For in the evolution of the early Universe, just such a stage was reached as it cooled, when, after about 700,000 years, the temperature dropped to around 3,000 K, and the entire Universe suddenly cleared. Before that, when it was hotter, the electrons and nuclei had not yet condensed into atoms, and the cosmos was filled with a bright opaque gas. When the temperature fell through 3,000 K, the electrons stuck to nuclei (mostly bare protons: hydrogen was the dominant component), light no longer interacted strongly with matter and so was no longer blocked by it, and transparency (which, at present-day temperatures, lets us see so far into the distance and the past), became universal. That moment when the temperature fell to 3,000 K was the moment when sight became possible. A thermal traveler *beginning* at 3,000 K and stepping down a power of ten would have been astonished at the rich complexity of the world that came into being in the relative thermal silence of 300 K.

At 3,000 K the remains of the picnic are lost to sight. At still higher temperatures, even the particles within nuclei (the *nucleons*) are liberated from the intense forces that bind them. By 30,000,000,000 K, about 2,000 times hotter than the center of the Sun, no nuclei can survive: so much energy is everywhere available that nucleons can separate from their neighbors and add to the chaos of the world. At ten times hotter still, even the protons and nuclei have been disrupted. Apart from major differences in densities (for the Universe is not being compressed as we heat the picnic), we have arrived at a condition similar to that of the early Universe about a hundredth of a second after its formation. The picnic party has been thoroughly burnt away.

Hotter Than Infinity

Although it might seem that our journey must stop as the temperature approaches infinity, we can go on, by a devious interpretation of what we mean by temperature. It is possible, in a certain sense, to achieve a hotness greater than that corresponding to infinite temperature. In order to see this, let us consider the model universe again, but now arranged so that more atoms are ON than OFF (see below). This arrangement cannot be reached naturally by thermal means, because all actual thermal reservoirs have fewer atoms ON than OFF; so the tendency of energy to disperse could never lead to a part of the universe to have more atoms ON than OFF. However, in our model (and in real life) we can contrive this artificial arrangement. Let us see what properties the system will then have.

A thermally isolated system can be put into a state of negative temperature if more of its atoms are ON than OFF. Here 80 are ON and 20 are OFF, and the temperature is −0.72.

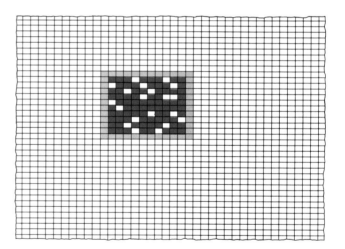

First, the temperature of the system shown above is *negative*. This follows from our interpretation of temperature as the logarithm of the ratio of numbers of atoms ON to OFF. In the arrangement shown, where 80 are ON and 20 are OFF,

$$Temperature = 1/\log\ (Number_{OFF}/Number_{ON})$$
$$= 1/\log\ (20/80) = 1/\log\ (0.25) = -0.72.$$

Hence, the temperature is −0.72 (in the dimensionless units we are using; this would correspond to −0.72 K only by accident, depending on how much energy was needed to turn the atoms ON: see Appendix 2).

Second, the system at a negative temperature possesses *more* energy than it does at any positive temperature. Therefore, in a certain sense, a system with a negative temperature is *hotter* than one with a positive temperature! In particular, it is even hotter than a system with an *infinite* positive temperature. Hence, we are now, by devious interpretation, beyond infinity.

Let us look more closely at the last remark. An infinite temperature corresponds to *equal populations* of ON-ness and OFF-ness. But to turn even one more atom ON requires energy, however we do it, and simultaneously takes us into the devious regime of negative temperatures. We have seen that having 80 atoms ON in a system of 100 corresponds to a temperature of −0.72; if we had only 51 ON (and 49 OFF) the temperature is even more negative, at −25. If we were considering a much bigger system, then we could approach ratios of ON to OFF that are very close to unity. For instance, if the system consisted of a million atoms, then an infinite temperature corresponds to 500,000 atoms ON and the same number OFF, and turning only one more ON sends the temperature plunging from infinity to −250,000.

If we extend this argument to real systems, which consist of billions of atoms, then one more atom being turned ON after infinite temperature has been attained sends the temperature down to virtually negatively infinite values. Then, as more atoms are turned ON as energy is forced in, the temperature *rises* toward zero, but always remains on the negative side of it. The complete course of changes for the 100-atom system is shown below.

Notice that we have wild behavior of the measurement we call temperature, whereas the population of atoms is changing smoothly through an ON : OFF ratio of unity. The reason for the wildness lies in the definition of

The temperature of a 100-atom system in the model universe depends on the number of atoms that are ON. When fewer than half are ON, the temperature is positive and finite (and is zero when none is ON). The temperature is infinite when exactly half are ON. One more ON sends the temperature plunging from infinity down to −25. (In a system of many more atoms, one ON beyond halfway would send the temperature down closer to minus infinity.)

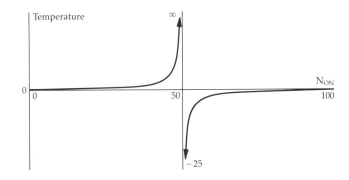

If temperature were defined as the negative reciprocal of the conventional temperature (that is, New temperature = −1/Old temperature*), temperatures would vary smoothly from minus infinity (what we now call absolute zero) through zero (for equal numbers of ON and OFF), and on to plus infinity when all atoms are ON.*

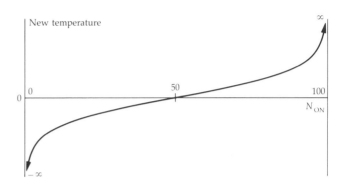

temperature as the reciprocal of the logarithm: if we redefine temperature as −1/*Temperature*, so that

$$New\ temperature = -1/Old\ temperature = \log\ (Number_{\mathrm{ON}}/Number_{\mathrm{OFF}}),$$

the wildness disappears. This is shown above: very cold (corresponding to an *Old temperature* of zero, and therefore to *New temperature* of minus infinity) lies infinitely low on the diagram; therefore no one should be surprised that "absolute zero" is unattainable! Supplying energy to the system raises the *New temperature* toward zero: equal numbers of ON and OFF then occur when *New temperature* is exactly zero (for that corresponds to an *Old temperature* of infinity). Then adding more energy takes the energy of the system smoothly higher as more atoms are turned ON, and the *New temperature* glides smoothly upward. When all the atoms are ON, the *New temperature* is infinite. How much simpler this definition of temperature would have been! Indeed, it is suggested naturally by Boltzmann's approach to thermodynamics; but the conventional *Old temperature*, which grew out of our familiarity with small ranges of temperatures around 300 K, is now too firmly ingrained for the conversion to the new to be possible (and perhaps even desirable).

What remains for us to explore is the working of the Second Law when systems have been artificially organized to have negative temperatures. Such systems can be prepared (although, of course, they cannot be in thermal equilibrium with real environments, and have to be protected in special ways). Some lasers, for instance, are systems described by negative temperatures. In a laser, atoms in a sample are first electronically excited (see figure on next page), then release their energy in unison, giving rise to an intense, coherent beam of light. The illustration on the next page is the analog of what we have already been considering, for more atoms are excited (are ON) than are not (are OFF). Hence the initial, prefiring state of

A laser beam can be regarded as being generated by a system with a negative temperature.

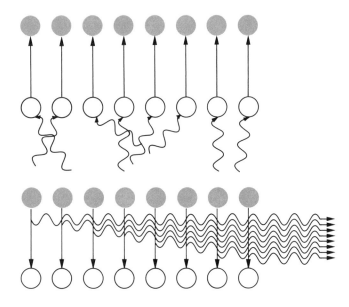

A laser works as follows. Light from an exciting source excites the atoms of the active medium into an upper state; that is, many atoms are turned ON, and the temperature becomes negative (if enough are turned ON). One photon is released as one of the atoms falls back into its lower-energy state. This photon stimulates another atom to generate a photon of the same frequency. These two photons stimulate a third atom to emit, and so on. As a result, there is a cascade of photons, and a bright ray of coherent radiation is generated.

the laser (of this type at least) is described by a negative temperature. We could also use special spectroscopic techniques to drag the majority of nuclear spins in a sample into orientations corresponding to their higher-energy state (see top figure on facing page); this too corresponds to a negative temperature. Since nuclear spins interact only very weakly with their environment, the nuclei can remain for long periods at negative temperatures, and energy will only very slowly jostle out into the surroundings.

Having seen that negative temperatures may be contrived, we can now consider the properties of steam engines (or their very sophisticated analogs) that might one day make use of reservoirs at negative temperature. Transferring heat *to* a reservoir with a negative temperature *reduces* the entropy of the reservoir; we arrive at this conclusion from the definition of entropy as *Heat supplied/Temperature,* but with the *Temperature* in the denominator given a negative value. As might be expected, this results in some bizarre effects.

The most important consequence of working at negative temperatures is that the Kelvin statement of the Second Law is contravened: heat can be completely converted into work, with no other change elsewhere. In order to see this vindication of Jack Rogue, consider the operation of the engine shown to the right.

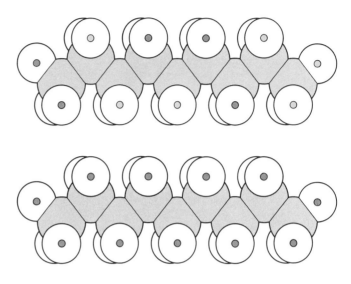

Under normal conditions the nuclei of atoms have a random arrangement of their spins. This is depicted for a hydrocarbon in the upper illustration: green denotes a proton (the nucleus of hydrogen) spin that is DOWN, yellow one that is UP. (Normal carbon nuclei do not spin.) However, the proton spins may be twisted into an arrangement in which all are DOWN (lower illustration). This corresponds to their being in the **upper** *energy state in a magnetic field. Therefore the nuclear spins are at a negative temperature (but the rest of the molecule is near room temperature).*

We take the cold reservoir to be the one with the less-negative temperature. Let its temperature be −2.5 on some scale. The hot reservoir, the one with a more negative temperature, is taken to be at −5.0. Then the heat engine can be imagined as operating as follows. First, heat is withdrawn from the hot reservoir. This *increases* the entropy of the hot reservoir by an amount (*Heat withdrawn*)/5.0. Some of this energy is converted into work, and the remainder is discarded into the cold reservoir. This discarded energy *reduces* the entropy of the cold reservoir by an amount (*Heat discarded*)/2.5. Now, suppose that *no* energy is discarded as heat. Then the total change of entropy of the Universe is still *positive*, because of the withdrawal of heat from the hot reservoir. This is a natural process, according to the entropy principle. Indeed, we may even throw away the cold reservoir: no heat need be discarded. The engine is anti-Kelvin!

This conclusion is so important that we should enquire more deeply into its feasibility, for now the entropy principle has apparently turned against the Second Law, the summary of experience it was introduced to capture. Consider the events involved in terms of the model universe in the figure on the next page. A reservoir at a temperature of −5.0 has 55 atoms ON out of every 100. With that distribution of ONs, there is an amount of entropy determined by how much the Demon can deploy them. If some of those ONs seep out, then the Demon has *more* scope. Indeed, we saw long ago (in the figure on page 70) that when more than half the atoms are ON, the Demon's freedom to deploy ONs increases as ONs turn

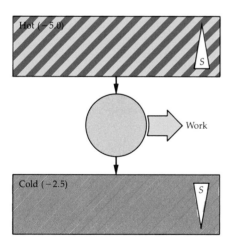

The changes of entropy that occur when heat is extracted from a hot source with a negative temperature and dumped in a cold source with a negative temperature.

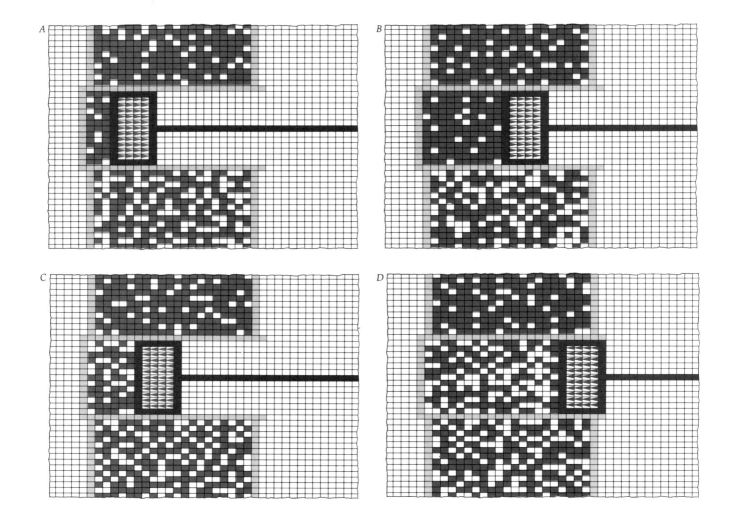

OFF; so reducing the number of ONs in the reservoir corresponds to an increase of chaos. At this microscopic level it is therefore entirely plausible that heat may be converted completely into work, for that continues to correspond to the triumph of chaos.

The extraordinary feature of this argument is that the Second Law has given birth to a more powerful child: the entropy principle appears to be more powerful than the law of experience on which it was founded. Of course, Kelvin had no experience with systems at negative temperatures, and so we may presume that the Second Law as he (and Clausius) ex-

The model of a Carnot engine working between reservoirs with negative temperatures. In going from A to B, atoms of the gas drive out the piston, but are maintained at the same temperature by thermal contact with the hot reservoir: the proportion of atoms ON remains constant during this power stroke. In the adiabatic stage from B to C, atoms turn OFF as work is done, and the ON:OFF ratio falls to that of the cold sink, but the number ON still exceeds the number OFF. The isothermal compression step from C to D requires work: atoms are turned ON, but the ON:OFF ratio is preserved by the thermal contact with the cold sink. In going from D to A the work done on the gas turns even more atoms ON, and the ON:OFF ratio rises to the value characteristic of the hot source.

pressed it is a report on only half the possibilities of experience. If systems of negative temperature had been available, there would have been two parts to the Second Law. Indeed, as far as I know, people still have had no direct experience with this "Through the Looking-Glass" domain of the conversion of heat into work; but we may presume that the behavior will be as we have described.

Toward Life

We have traveled into the dark quiet of cold, where physics almost ends, and into the searing heat where matter blends with radiation. We have seen that we can go into a new domain of thermodynamics, where negative temperatures promise to extend our experience. We have even seen that chaos may be harnessed and used to drive systems against nature, that noise may be harnessed to achieve silence. Next we shall explore the full flourishing of the constructive power of chaos. We shall see how chaos can run apparently against Nature, and achieve that most unnatural of ends, life itself.

On the left is a fragment of a system with a strongly negative temperature; there is little opportunity for redeployment of ONs, because there are few gaps. On the right is the same system after some ONs have escaped: now there is more freedom for the deployment of ONs, and so the entropy is greater, even though the energy present is less than on the left.

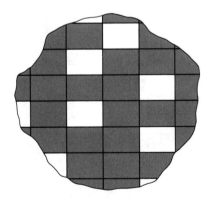
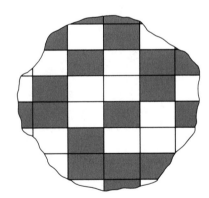

8 CONSTRUCTIVE CHAOS

The unnatural may be contrived at the expense of the natural. So long as we can drive one change by another, one change may be *constructive* and lead to a local reduction of entropy (such as in cooling something to below the temperature of its surroundings). But elsewhere, and coupled to the first, there must be a process that generates at least a compensating amount of entropy. This might take the form of the combustion of fuel in the electricity-generating station coupled to the refrigerator we are using to bring about cooling. There may be *local abatements* of chaos, which appear to us as the emergence of structure, but elsewhere there must be generated at least a compensating amount of chaos.

We have seen simple examples of the emergence of structure. For instance, the cooling of a substance reduces its thermal motion and endows it with order. We shall see that the complex, intricate structures characteristic of life itself may be produced if they are driven by the collapse elsewhere of the Universe into compensating chaos. This is where chaos emerges from its chrysalis. No longer the caterpillar of purposeless corruption, it now emerges as a constructional principle of such power that it inspires consciousness itself. Here at last the steam engine comes into its own: we are on the threshold of the apotheosis of corruption.

The Emergence of Intricate Structure

We begin (as always) with a primitive question, for Nature's primitive aspects lead us to its heart. Here we ask why a blob of oil does not disperse and dissolve in water. Why doesn't oil disperse like a drop of ink? What is the process that interferes with the tendency of matter to disperse?

The tendency to chaos is the constraint! Yet again, chaos deceives: what *appears* to be a lack of dispersal is in fact dispersal disguised.

First, it seems obvious (but will shortly prove to be quite wrong) that the spreading of the oil molecules of the blob through the surrounding water (see figure on next page) corresponds to the dispersal of particles. Correspondingly, we may also conclude that the spreading contributes

To the superficial observer, the breaking up and spreading of a blob of oil appears to be a dispersal, and thus appears to be a natural process.

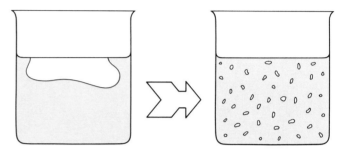

entropy to the Universe, for after the spreading the world is more chaotic. We shall leave it at that for the moment, because it is interesting to play a game of gradual discovery and see the importance of circumspection.

The dispersal of particles may be accompanied by a dispersal of energy; so there may be a second contribution to the overall change of entropy of the Universe. In order to decide whether this dispersal of energy corresponds to an increase or a decrease of entropy, we need to know whether energy is absorbed or released as the blob undergoes its hypothetical dispersal. In some situations energy would need to flow in as heat to allow the oil to dissolve, because the oil molecules are in an energetically less favorable environment when surrounded by water molecules than when surrounded by their own kind; so in these situations the entropy of the world outside the container is reduced, for the thermal motion out there is being quenched. In other situations, energy is released as the oil dissolves, and the hypothetically dissolving oil is accompanied by the heating of the surroundings. This happens if the molecules of oil disperse into an energetically more congenial environment than before. In these situations the entropy of the outside world is increased, because its thermal motion has been stimulated.

The second possibility is more difficult to accommodate than the first, because we then appear to have *two* positive contributions to the entropy; yet still the oil is observed not to dissolve. Since the entropy of the system is increasing with the oil's dispersal, and the entropy of the surroundings is also increasing because energy is escaping into them, it appears that the spontaneous direction of change is toward dissolution of the oil, for that is the direction of increasing overall entropy. But oil, even the kind of oil that would release energy if it were to dissolve, does not in fact dissolve. What have we forgotten?

We have forgotten the invisible background of water. As so often happens, the easily neglected, the commonplace, plays the dominating role.

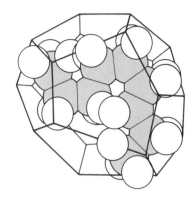

When a hydrocarbon is surrounded by water, the water molecules form a cage around it; so the water needs to acquire more order than before.

Consider the dispersal in more detail, and think about what happens to the water as molecules of oil float away from the blob. As the molecules disperse, they are surrounded by water molecules (as in figure on left), and each one sits in a delicate molecular cage. These cages are *structures*. Their existence means that the water molecules become *more organized* when oil molecules are among them. In these invisible cages, we have a contribution to the order of the world.

The invisible contribution of the commonplace is the key to the experimental fact that oil does not dissolve in water. The reduction of disorder that accompanies the formation of the cages corresponds to a significant decrease in entropy. This decrease overcomes not only the increase arising from the physical dispersal of the particles, but also the increase in entropy of the surroundings if energy is released into them. Hence the natural, spontaneous direction of change is from dispersed oil molecules to blob (see figure below); the dispersal of the blob is unnatural. In order to bring about its dispersal, we need to do work, such as beating and whisking: the production of some sauces in cooking is just one example of the power of the Second Law.

To the circumspect observer, the congregation of oil droplets and molecules into a single blob is the process that corresponds to dispersal!

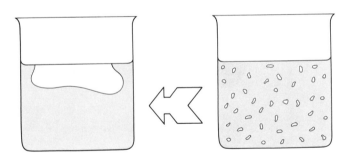

Note that a Martini is not a greasy concoction. This is also an aspect of the Second Law, but now the Law is working in the direction of clarity, not cloudiness. Suppose we were to adopt a particularly savage, fundamentalist way of mixing a Martini, and pour pure alcohol (ethanol, CH_3CH_2OH in chemical composition) into water. If the alcohol molecules behaved exactly like oil molecules, they would congregate into a blob, and the mixture would be oily. But alcohol molecules are not exactly the same as oil. Although they have a tiny hydrocarbon backbone (oil molecules are only backbone, having around ten carbon atoms in a chain), they do possess an

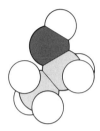

An ethanol molecule is like a hydrocarbon that has been singed with oxygen at one end; so at that end it resembles water.

oxygen atom too (the red sphere in the figure to the left). To that extent they also resemble water molecules quite closely. Water molecules stick together by virtue of a *hydrogen bond,* in which the electron-starved hydrogen atom lies between electron-rich oxygen atoms belonging to different molecules, to give structures such as

$$ \begin{array}{ccc} H & & H \\ \diagdown & & \diagdown \\ O\!-\!H\!\cdots\!O & \\ & & \diagdown \\ & & H \end{array} $$

This sort of link occurs between the alcohol and the surrounding water. Since the oxygen end of the alcohol molecule looks more like water than any part of an oil molecule does, it sits more readily among the existing structure of the surrounding water, and water cages need not be constructed to the same extent. Hence there is less increase of order in the structure of the water when alcohol spreads into it. Overall, the balance of changes in order now lies in favor of dissolution. Hence the alcohol mixes with the water, and the Martini is clear. Next time you gaze into a Martini, reflect on the victory won in your favor by the Second Law.

The process of *apparently* running against chaos when oil congregates into a blob is called the *hydrophobic effect.* We can find its analog in the molecules that govern the processes of life, the *proteins* of living systems. We shall now see how chaos results in the construction of these exquisitely ordered forms.

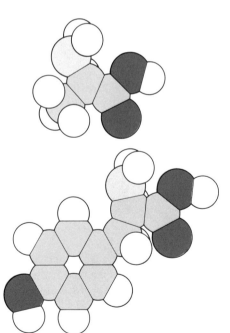

Two amino acids. The smaller molecule is ala-nine *(Ala), and the larger is* tyrosine *(Tyr). Blue is nitrogen, red oxygen.*

Proteins

A protein is a string of amino acids: these are quite small organic molecules, with the common feature that they contain the grouping of carbon, oxygen, and nitrogen atoms shown to the left. (The same carbon atom need not carry the N atom and the —COOH group, but it does in the acids that make up proteins, and we shall deal only with them.) They get their name from the fact that the —NH$_2$ group is called an *amino* group in organic chemistry, and the —COOH group causes compounds that contain it to behave like acids. There are twenty naturally occurring amino acids, but since they can link together in any combination, the number of protein molecules they can form is endless. Twelve of the acids can be synthesized by the human body, but the remaining eight have to be included in the diet, and are therefore called the *essential* acids.

A single peptide link in a polypeptide protein chain. This has the structure —CO—NH—, and occurs between each neighboring peptide (amino acid) unit along the chain.

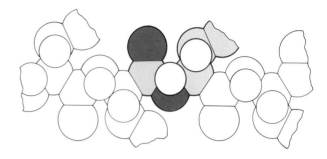

The bond between two neighboring amino acids in a protein molecule is called a *peptide* link, and its form is illustrated above. The *primary* structure of a protein is stated by describing the sequence of amino acids that it contains: several hundred may sometimes occur in a chain. The chains from which hemoglobin are formed, for instance, each contain around 145 amino-acid groups, and each molecule contains four such chains. The immense complexity of proteins can begin to be appreciated.

The chain of amino-acid groups called the *peptide chain*, coils itself into a helix, the *alpha* helix, under the influence principally of the interactions between the oxygen and the nitrogen atoms of the peptide links. Just as water molecules are held to each other by hydrogen bonds, so the hydrogen attached to the nitrogen of an amino group can stick it to an oxygen atom of a —COOH group, as in the figure on the next page. The link $>$N—H····O— so formed is quite similar to the —O—H····O— link between water molecules. It is strong enough to establish a spiral structure on the rhythmical array of peptide bonds, and so the alpha helix is coiled into existence.

We now have our first example of the constructive power of chaos in biochemically important molecules. (Another is lurking in the formation of the chain itself, but to that we shall return.) The alpha helix is favored over

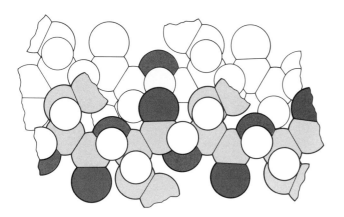

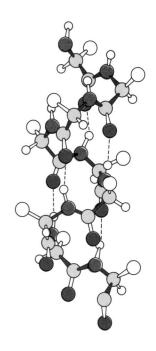

A hydrogen bond, —N—H····O—C—, *may
occur between peptide links at different locations
in the peptide chain (above: the linking hydrogen has been picked out in yellow). As a result
of many of these bonds, the chain is twisted
into a helix, the alpha helix (right).*

the random bundle because it corresponds to the *more* chaotic form of the
Universe. The chain itself is certainly *less* chaotic on account of the more
ordered spiral disposition of the peptide links, but the world is *more* chaotic
on account of the energy released into the environment when the strong
hydrogen bonds form. The competition lies in favor of the dispersal of
energy, not in favor of the chaos of the disordered chain. The world is
more disorganized overall when the helix is formed than when it is not.

Most proteins are not in the form of lengthy cylinders of alpha helix:
the figure to the right shows that the four strands that make up a molecule
of hemoglobin are bent and twisted, a bit like a snarled-up and discarded
drinking straw. This apparently chaotic collapse, however, is yet one more
delusion: the snarled-up straw is, in fact, a precisely sculpted artifact of
chaos. In the strand we are seeing not its random collapse into chaos, but
the emergence of a precisely ordered form. Although one hemoglobin
strand may look like an untidy shambles of atoms, the exact repetition of
that shape in billions of molecules tells us that it is ordered. A billion
snarled-up drinking straws are all different; they are disordered. A billion
hemoglobin molecules are identical; they are ordered.

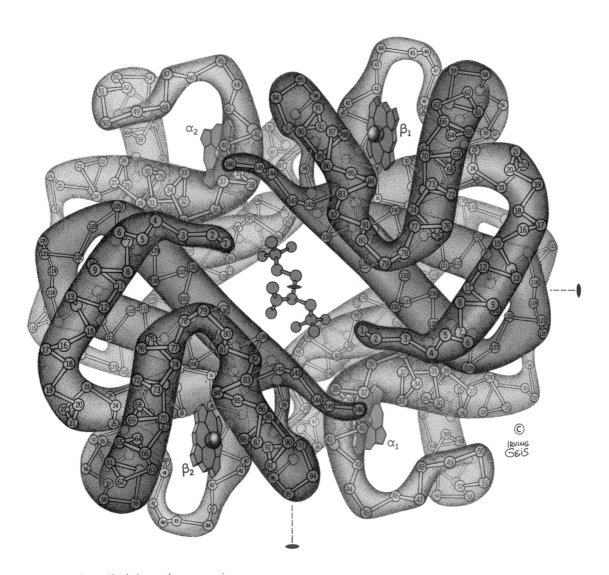

The four polypeptide chains, each one wrapping round an iron atom, that stack together to form the molecule of hemoglobin.

The *tertiary structure* of a protein molecule, the precise manner in which the alpha helix is crumpled, bent, and twisted into its specific functional shape, is determined by interactions between different parts of the chain. One of the most important interactions is hydrogen bonding between peptide groups in different parts of the helix. These bonds can buckle and strap it into its bent and crumpled form. Another is the interaction of electrical charges on different parts of the helix: like charges will push each other apart; opposite charges will pull each other together. When the helix bends in response, energy is released and goes into the surroundings. The buckled structure is then frozen into shape, because energy will probably not jostle back into the helix from its wide dispersal outside—unless there is a lot of energy around, as when an egg is boiled. Boiling an egg provides enough energy to the vibration of the atoms of the helix for the tertiary and secondary structures to be broken down. At a molecular level, boiling an egg is the unwinding and unzipping of a crumpled helix.

But the interaction we shall concentrate on is the hydrophobic bond, for it plays a crucial role in bringing about and sustaining the tertiary structure of proteins, and therefore organizing the fine-tuning of their shapes that renders them fit to work. This oil-blob-forming interaction plays a role in protein formation because many of the amino acids in proteins have bulky hydrocarbon parts, and to that extent behave like molecules of oil. By the same argument as before, water molecules would need to form cage-like structures if these oily parts were projecting among them, and these structures would represent a large reduction of entropy; so the natural direction of change is for these oily parts to become buried inside the protein, and away from contact with the surrounding water. If they burrow inward, then their less obtrusive (more water-like) parts stick out into the water, and these demand less structure formation from the water molecules. Thus the protein helix is driven into a bundle in which the water-like ("hydrophilic") parts of the peptide groups are on the surface and the hydrocarbon ("hydrophobic") parts are inside. Entropy thus crumples the helix, and the delicate balance of locations of hydrophobic and hydrophilic parts of the chain causes it to fold to a precise, seemingly architected, form.

The constructive effect of entropy may also be the reason why four crumpled strands of proteins stick together to form the complete molecule of hemoglobin. (This agglomeration of strands to form a complete protein is its *quaternary structure*.) When the four strands, each with its characteristic tertiary structure, come together, the entropy of the world is increased, partly because of the energy that is released (though that may not be very great), but also (and perhaps mainly) because the hydrophobic parts of each strand shelter each other. They group together like oil droplet sticking to oil droplet, and so the surrounding water molecules no longer need to form a receptive cage. Thus the tendency of the world to chaos squeezes the strands together: it builds the quaternary structure at the expense of disorder in the surrounding medium.

Thus we see that the tendency to chaos builds both the quaternary and the tertiary structures of proteins, for the ordered helices bend and buckle, and the surroundings become more disorganized. It also builds their secondary structures, for it is through the consequences for entropy that the hydrogen bonds, on which the structure depends, are locked into stability, and wind the ordered helix from the random chain of links.

But what of the *primary* structure, the ordering of the amino acids themselves into the original chain? A scattering of disconnected amino-acid molecules has a much higher entropy than when they are pinned together into a geometrically chaotic but still specific sequence. Does a Director have to direct this stage of the emergence of living things, or can this essential primary step also emerge as a consequence of chaos?

Free Energy

If only energy *were* free! The primary backbone of proteins is built when amino atoms are joined together by a sequence of chemical reactions. These reactions are very complex, and depend on the action of *enzymes*. Enzymes are varieties of proteins that act like machines on a production line: they stick a group of atoms on here, snip others off there, and hand the modified molecule on to the next machine in line. Nevertheless, *overall* the reaction is very simple: two amino acids are joined together by a newly formed peptide link, and a water molecule is eliminated (see figure on next page). *Which* amino acids are joined depends on the template present in the cell: that template is DNA, the most sophisticated molecule of all (look back at the figure on page 7). Indeed, all the remarks that we shall make about biologically important reactions in speaking of proteins apply (give or take a myriad of details) to DNA too.

The process by which two amino acids are joined, and by which the chain is successively and precisely extended according to the recipe sent to the site of synthesis by the cell's DNA, is *overall* a simple chemical reaction. All the arguments we have discussed so far about reactions, and about how they result from the dispersal of energy, are applicable. Yet there seems to be a problem. When the ordered chain is built from scattered amino acids, there appears to be a *reduction* of the entropy. However, we have seen that apparent reductions are often only local abatements, and that, if we look hard, we can always find a more-disordered region of the world elsewhere. We must therefore explore the chain of events that gener-

The net result of the reactions that produce a peptide link is very simple: two hydrogen atoms and one oxygen atom are eliminated (as water) from the amino acids being linked. The individual steps that lead to this result are complex, but we need consider only the overall outcome.

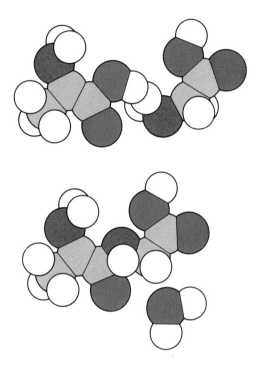

ates proteins: we shall see why we need to eat, and we shall see why, when we fail to eat, we die. In short, we need to be *sustained*. This is a rich concept, and it will take us far beyond the immediate application that has stimulated its introduction here.

We shall approach the question of why we need to eat in a fairly circuitous way, but it is a route that will give us some important insights into the workings of the world. In the first place we shall explore the criterion that decides whether one chemical reaction can be harnessed to another and used to drive the latter in an unnatural direction. This is the exact analog in chemistry of processes in the everyday world, where a more vigorous spontaneous process is used to drive another in an unnatural direction (such as a refrigerator harnessed to an exploding nucleus or to a stream of falling water, or even an actual cart harnessed to an actual well-fed horse).

The central lesson of the Second Law is that natural processes are accompanied by an increase in the entropy of the Universe. A coupling of two processes may cause one of them to go in an unnatural direction if enough chaos is generated by the other to increase the chaos of the world overall.

Two unconnected weights each tend to fall downward; however, if they are linked, the heavier one will raise the lighter. This is an analog of the behavior of chemical reactions.

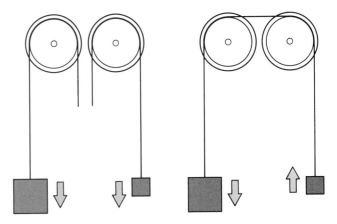

This is like the two weights shown above. Each one alone tends to fall to the ground, but, when they are coupled, the heavier raises the lighter, driving it in an unnatural direction. How, though, do we *assess* the driving power of a reaction? If we can answer that, and can set up a scale of driving powers, then we could easily assess whether any given reaction (such as the synthesis of a protein) can be driven by another (such as the digestion of food), just as we can decide whether one weight can lift another.

Consider a reaction that liberates energy as heat (such reactions are called *exothermic*). Suppose the reaction also reduces the entropy of the system itself. For instance, this is true for the oxidation of metallic iron: we saw on p. 111 that the reaction liberates heat, but reduces the entropy of the substances overall (largely because the large volume of gaseous oxygen collapses into the tiny heap of oxide). Suppose, furthermore, that we want to harness the energy that the reaction produces, not merely to heat the world, but to do work in it. For instance, we might be burning iron in a furnace and using the energy to drive some kind of vehicle (burning coal would be a more familiar example). Since transporting the energy released by the reaction to the outside world as (quasistatic) *work* does not change the entropy of the surroundings, we are now confronted with an overall *decrease* of the Universe's entropy, because the reaction substances undergo a reduction of entropy, but there is no change in the surroundings. It follows that the conversion of *all* the energy released by this type of reaction into work is not a natural process. Note carefully the following distinction. All the energy released by a reaction may emerge into the surroundings as heat, for that increases their entropy; not all the energy released may emerge into the surroundings as work, for if it did the overall change of entropy would be negative, and the Universe would have shifted spontaneously to a less probable state.

Although not all the energy released by a reaction is available for doing work, perhaps if we allow *some* of the energy to escape as heat, enough entropy may be generated in the surroundings for the process to be spontaneous, even though we withdraw the remainder of the change of energy as work. We can then ask the following question: what is the *minimum* amount of energy that must leak into the surroundings as heat in order to generate enough entropy there to allow the reaction to be spontaneous?

The changes of entropy that occur in the course of a chemical reaction that can exchange heat with its surroundings. If the entropy of the system falls, some energy must be released as heat if the reaction is to be spontaneous; so not all the energy released is available for doing work.

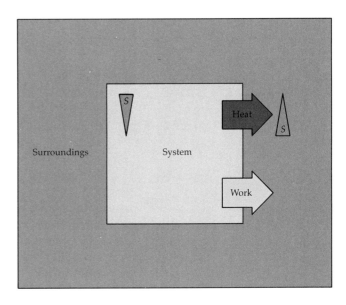

Suppose the reaction reduces the entropy of the system by an amount *Entropy change* (see figure above). In order for the reaction to proceed spontaneously, at least this amount of entropy must be generated in the surroundings. But we have seen that the entropy generated there is always given by the expression (*Heat supplied*)/*Temperature*. Therefore the minimum amount of energy that must be supplied as heat to the surroundings by the exothermic reaction is obtained by equating these two expressions and solving for *Heat supplied*. Clearly, the minimum heat that must be released to the surroundings is the product of the temperature and the reduction of entropy:

Josiah Willard Gibbs (1839–1903).

Minimum heating = Temperature × (Entropy change).

It follows that the energy *not* available for doing work when the reaction occurs is equal to the expression on the right. This is normally written symbolically as $T\Delta S$. On the other hand, the energy that *is* available for doing work is the difference between the total energy released and the amount we have just calculated. In other words, the *free energy*, the energy available for doing work, is given by

Free energy = (Total energy) − Temperature × (Entropy change).

The free energy* of a reaction is its single most important thermodynamic property, and it will now stand at the center of our stage, just as the man who introduced it, Josiah Gibbs, is the single most important contributor to chemical thermodynamics.

Josiah Willard Gibbs was born a member of our third wave of minds, the generation of Boltzmann, the successors of Kelvin, Joule, and Clausius. Although those three had established the formalism of the subject, and Boltzmann deserves the credit for peering inside it for explanations, Gibbs is responsible for most extending its range. Working at Yale, where he spent most of his life, he developed a way of applying thermodynamics to chemistry. This "greatest of Americans, judged by his rank in science" thereby converted physical chemistry into a deductive science. Gibbs is the intellectual link between the steam engine and chemical reactions, and his obscurely published paper "On the equilibrium of heterogeneous substances" ranks with the various *Reflections* in its scientific impact (however, his reticent nature contributed to the slowness with which its greatness was recognized, especially in the United States). The span of Gibbs's interests is reflected in his life: he took his doctor's degree in engineering (the first such degree in the country) with a thesis on gearing, and while developing with great subtlety and elegance of thought what the theoretical implications of thermodynamics were for chemistry (which some had considered beyond the limits of rational explanation), he maintained his interest in practical matters. It is reported, for instance, that he prescribed and ground the lenses for his own spectacles.

So far we have been considering a reaction in which the entropy of the system decreased; heat must enter the surroundings in order for such a

---

* The quantity normally considered by the chemist is the *Gibbs free energy:* it relates to changes taking place when the pressure is constant. A slightly different property, the *Helmholtz free energy*, arises when the changes are taking place at constant volume. We are not troubling to make the distinction here, just as we are not distinguishing between internal energy and enthalpy. These points are taken up in Appendix 2.

reaction to be spontaneous. But suppose the entropy of the reaction system were to *increase* during the reaction. Now something rather interesting can happen, because we may then still have no net decrease in the entropy of the Universe even if the entropy of the surroundings is reduced. We can then afford to allow energy to flow into the system as heat from the surroundings (so reducing their entropy) and to emerge back into the outside world as zero-entropy work. That is, if the reaction releases energy, we can draw off *all* that energy as work, and we can draw off as work the energy that was sucked into the system from the surroundings as heat. Hence we can get *more* energy out of the reaction as work than we could as heat alone! Reactions such as these are *energy processors*, and convert low-quality energy of the surroundings into useful work.

The importance of the free energy arises from two interpretations of its significance. First, there is the one we have just seen. It tells us the maximum amount of work we can expect to obtain from a given chemical reaction that is taking place in thermal contact with the outside world. This work may be significantly different from the amount of energy the reaction can liberate as heat, because the reaction can act as an energy processor (see figure below). Therefore, when we are interested in the generation of *ordered* motion, we should talk in terms of the free energy of a process. (The free energy change in the course of a reaction can be calculated by using tables of energies and entropies of substances; so it is based on tempera-

A reaction that increases the entropy of the system can act as an energy processor: it can draw in energy as heat, and add it as work to the energy it is already releasing. Although the entropy of the surroundings drops, this is compensated for by the increase of entropy of the system.

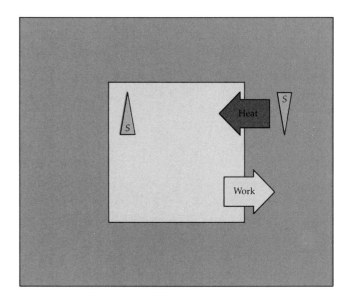

ture measurements, just like any other thermodynamic property.) The work available may be less or more than the heat that is available. A reaction from which the available work is greater than the available heat is, of course, not generating energy. All we are saying is that enough chaos is generated in the system itself to allow the thermal motion of the surroundings to be converted into ordered motion, the Universe still becoming more disordered overall. The *practical* importance of the free energy is therefore that its value tells us how much energy a chemical reaction can deliver as, for instance, electrical work. This is especially important when the reaction is the core of a battery or a fuel cell.

The second interpretation of the free energy, and one that is full of significance for our present discussion, concerns the spontaneity of the reaction. We have stressed previously that processes capable of doing work must be *natural* processes. It would be absurd, for instance, to try to use an engine to do work if it has to be driven by another more powerful engine! We can turn this idea round. If a process can do work, then it must be spontaneous. Therefore the change of free energy accompanying a reaction tells us whether the reaction is spontaneous or not. In particular, *if the free energy decreases in the course of a reaction, then the reaction is spontaneous in that direction.** In simple terms, the tendency of chemical reactions is *downward* in free energy (see figure on left).

Reactions fall to lower free energy just as initially stationary particles fall to lower potential energy. Here lies a deep analogy between dynamics and thermodynamics. However, there is a crucial difference that must be borne in mind. We have constantly stressed that thermodynamic systems do *not* tend toward states of lower energy. Therefore the tendency to fall to lower free energy must not be interpreted literally in terms of falling down in energy. *The Universe falls upward in entropy:* that is the only law of spontaneous change. The free energy is, in fact, just a disguised form of the total entropy of the Universe, even though we have introduced it in a way that might conceal the connection, and even though it carries the name ''energy.''

We opened this section by drawing an analogy between two weights and a chemical reaction. Now the analogy can be expressed precisely. A chemical reaction is like a weight, but a weight that falls down in *free* energy, not potential energy. The analogy with a heavy weight raising a

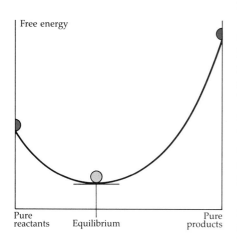

Chemical reactions taking place under conditions of constant pressure and temperature have a spontaneous tendency to lower free energy. Either of the two starting compositions of a reaction mixture (with free energies represented by the red blobs) will tend to the equilibrium composition (yellow blob).

---

* The identification of the signal of spontaneity with a *negative* change of free energy is very clear in the formal, mathematical development of thermodynamics (Appendix 2). Here let us simply accept that when work is done by a system, there is a reduction of its energy, and therefore of its free energy too.

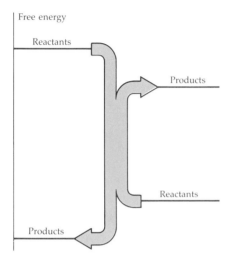

Free energy

Reactants

Products

Reactants

Products

A reaction that corresponds to a large decrease of free energy (on the left) can drive another reaction in an unnatural direction if coupled to it, like a heavy weight raising a lighter.

lighter one when they are connected in some way is that a chemical reaction with the greater fall through free energy may drive another in its unnatural direction if it is coupled to it (see figure on left).

This is the key to understanding biosynthesis. In the cells of living things, many reactions are coupled to each other. Although one reaction may have a tendency to run in the wrong direction (as a chain of amino acids may have a tendency to decompose, not form), it may be *forced* to run in the direction required if it is coupled to another reaction with a greater negative change of free energy. Some reactions are like light weights; others are the heavy weights.

The Unnatural Reactions of Life

Bodies and cells are like immensely complicated assemblies of gears. Here a heavy weight falls down the hill of free energy, and there, on account of the intricacy of the biochemical gearing that runs through the cell, it may drive a lighter weight up the hill of free energy, but not quite as far. In other words, the body sinks into corruption, but in such a remarkably interconnected way that in its progress downhill it gives rise to all the intricacies of life and consciousness. That is why we eat: we import material high up on the slope of free energy (in other words, material low in entropy, energy high in quality), and allow it to decay.* As food falls down the slope of free energy, and in due course becomes excrement, so our inner gears turn, and we take life.

In this chapter our goal has been to understand the processes that give rise to proteins, the worker bees of our hive of cells. Now we need to look at them more closely, and to watch the flow of energy, particularly the degradation of the quality of energy, that drives them into life. At each stage of the complex web of reactions that takes food from ingestion to excretion, and that simultaneously drives material into ever more complex forms elsewhere in the body, we shall see the Universe sinking just a little more into chaos. At each stage the Universe sinks into a state that is more probable than the last. Each moment confines the Demon more securely into the future, and makes it more and more improbable that it will ever reassemble the past.

_____

* That high quality emerges initially from the Sun, whose temperature is so high that its energy is stored with very low entropy. Energy of very excellent quality rains down on us daily, and is captured at its peak by photosynthesis. It then begins its progress through plants, and continues on through animals.

We shall take a very simplified run through the processes that power the body in order to catch a glimpse of the gearing that drive us. The subject of *bioenergetics* is immensely complicated, and in this primitive survey we cannot hope to capture its nuances. (The books on this subject that I have included in the bibliography will show you what corners we are now cutting, and enable you to explore the current state of knowledge of this remarkable field.)

The molecule on which we focus in this section is *adenosine triphosphate*, or ATP. It is a medium-sized molecule, as biochemical molecules go, and its structure is depicted in the figure below. Its crucial role arises from its ability to shed and reform the terminal *phosphate group*, the cluster of oxygens around the phosphorus atom at the end of its tail. ATP's terminal phosphate gets attached to the rest of the molecule (which is called *adenosine diphosphate*, ADP) under the influence of the oxidation of glucose, which in turn results from the breakdown of carbohydrates we ingest as food. This is our first example of a smaller weight being raised by a heavier one. The newly formed ATP molecule then moves to another reaction site, where its terminal phosphate breaks off, releasing its energy and driving another chemical reaction. This may be the construction of a peptide link leading to the formation of a protein, or a process in the nervous system that leads to the formation of an opinion. ATP is the crucial energetic intermediary, the gear wheel of life.

Glucose is burnt to carbon dioxide and water. This is the overall process of respiration and digestion. Of course, glucose does not burn with a flame inside us: bodies are much more subtle than hearths, and extract energy by means of chemical gear wheels, not by letting it idly burn to

A molecule of adenosine triphosphate *(ATP). The yellow atoms are phosphorus: when ATP forms ADP by losing the terminal phosphate (PO$_4$) group, it releases energy for use in another reaction.*

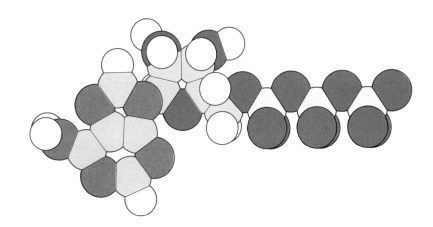

waste. When each molecule of glucose is destroyed, enough energy is released as the reaction rolls down the hill of free energy to produce about 93 molecules of ATP by attaching phosphate groups to ADP (but even Nature isn't perfect, and not all 93 molecules are produced).

The first step in the chain of reactions is *glycolysis*, where the glucose molecule is snipped in two. This drops down in free energy far enough to produce two ATP molecules. Note that the snipping apart of the glucose molecule is releasing energy, and structure is degrading; hence the collapse of the world into chaos can proceed, even though some of the energy is used to *construct* ATP molecules. The details of the processes involved in linking the degradation of glucose to the construction of ATP depend on the intricacies of enzymes, but we shall not look at them any further.

The first stage of extracting energy from carbohydrates ingested as food (after they have been broken down into glucose) is for the glucose (the molecule on the left) to break apart into the two pyruvate ions (CH₃.CO.COO⁻) shown on the right. This is the glycolysis *reaction.*

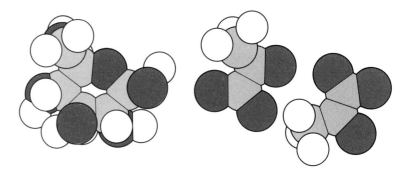

The disruption of glucose into its fragments (the *pyruvate ions* depicted above) gives up only a fraction of the energy it carries. The *mitochondria* of cells have evolved to make use of the remaining free energy. (These bacteria-sized organelles may originally have been bacteria that invaded and then colonized cells.) Note how the extraction of energy in a cell takes place in a different region from its application: the energy is extracted in the mitochondria and used elsewhere. It may be that Nature is carefully avoiding short-circuiting the imported energy. If the energy were extracted and employed in the same place, it might behave like a hot block of metal put in direct contact with a cold one: it might degrade directly, and the extraction of useful work might not take place.

The mitochondria are like *electrochemical cells*, the cells that are linked together in electric batteries and do electrical work (pushing electrons through a circuit) at the expense of a chemical reaction. In order to appreci-

The drawing on the left shows a typical (but simplified) version of an animal cell: the mitochondria (shown in more detail on the right) are the small organelles scattered through the cell. They act as its power supply.

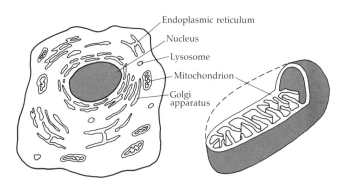

Endoplasmic reticulum
Nucleus
Lysosome
Mitochondrion
Golgi apparatus

ate the relation (and to see how the Second Law captures the chemical generation of electric power) we shall make another brief but important detour.

The Electrochemistry of Life

The idea behind the operation of an electrochemical cell is as follows. Suppose we were simply to dump a pellet of iron into a solution of copper sulfate. We would notice that copper is deposited on the surface of the iron, and that the pellet gradually crumbles and goes into solution. The overall process is that the copper ions in the solution capture electrons from the iron and so turn into neutral copper atoms, which deposit, while the iron atoms in the pellet that lose electrons become iron ions, which dissolve. We can think of this overall process as occurring in two steps: one is the capturing of electrons by the copper ions; the other is the loss of electrons by the iron. The process is spontaneous because it increases the entropy of the world (here largely because of the energy that jostles away to heat the surroundings). Put another way, it corresponds to a reduction of free energy (because there is an increase in the overall entropy of the Universe).

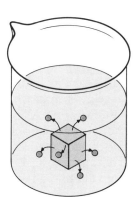

If a lump of iron is dropped into a copper sulfate solution, electrons are exchanged in random directions between the lump and the ions in the solution. There is no net flow of current, and the copper deposits on the surface of the iron, which gradually dissolves.

As the reaction is depicted to the left here, the transfer of electrons between the iron and the copper ions is higgledy-piggledy, and we cannot discern a *direction* of flow (apart from knowing that it is from iron to copper). But suppose we could find a way to allow the iron to dump its electrons into an electrode, to travel to another electrode, and there to be captured by the copper ions in the vicinity. Then the overall effect would be the same (iron would dissolve, and copper would deposit), but we would

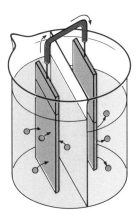

*If the copper and iron are in separate compart-
ments, then the tendency of the copper to deposit and the iron to dissolve results in an orderly, coherent flow of electrons through the
wire joining them. Hence, the chemical reaction
generates an electric current (which can be used
to do work).*

have achieved a flow of electrons. In other words, *the chemical reaction would
have established an electric current.* The imposition of this orderliness is the
basis of the operation of a cell (see figure to left).

An electrochemical cell is simply a device in which the processes of
electron loss and capture are separated. As the Universe sinks into chaos,
the electrons have to migrate through whatever external circuitry is provided. The chemical reaction establishes an ordered flow of current, and it
is this ordered flow that we use to run electric motors attached to the
circuit.

This raises an important point: how much work can we extract from a
cell? As we have explained, since at least a little energy must be discarded
as heat, the quantity of work available is equal to the change of free energy
during the cell reaction. Hence, knowing the free energy of the reaction
(which we can often look up in tables) enables us to predict the electrical
energy a cell can produce.

Now we can return to Nature's electrochemical cell, the mitochondrion. The processes that take place in mitochondria are a sequence of
transfers of electrons from molecule to molecule. Overall, they extract electrons from the pyruvate ion, the ash of glycolysis, and allow them to be
captured by oxygen. The outcome is that the pyruvate ion falls apart into
carbon dioxide, the oxygen ions pick up hydrogen ions and emerge as
water, and the flow of electrons constitutes an orderly process that can be
used to do work (see upper figure on facing page). In particular, the flow
may be used to build ATP molecules from ADP.

The steps inside the mitochondria constitute the *terminal respiratory
chain.* Needless to say, each one is very complicated. Nevertheless, the
pathway is quite well-understood, and can be compared to a series of electrochemical cells. The electron deposited by the pyruvate ion (and that
dumping itself involves a whole complex sequence of events known as the
Krebs cycle) is lowered gently down a staircase of cells, each of which has a
little electrically powered production line for manufacturing ATP from the
ADP that is lying around. This staircase is shown schematically in the
figure to the right, and the production capacity of each step is marked.
When the electron has reached the lowest level, and is attached to the
oxygen we breathed in (and which has been brought to the cell wrapped in
another protein, hemoglobin), its work is done. Lying in its wake is a
handful of ATP molecules (38, to be precise, for each molecule of glucose
consumed). The entropy of the world is now greater than it was before, but
in the litter of chaos there now lie more structured species, the molecules of
ATP.

At this point ATP is "charged": its energy is ON. If it now lets go of its
newly acquired phosphate group, and if the letting go is geared through

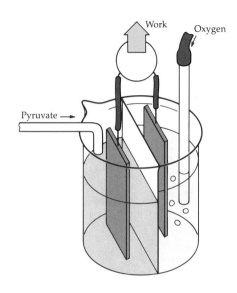

The terminal respiratory chain *involves the oxidation of the pyruvate ion (produced by glycolysis) to carbon dioxide and water. The overall reaction transfers electrons to the inhaled oxygen, and if the transfer is orderly (as in a cell), it can be used for doing work, such as building ATP, proteins, or DNA.*

enzymes to other reactions, then the store of free energy may be used to raise other, less-weighty chemical reactions. Each peptide link represents an *increase* of free energy, partly because the resulting structure is so much more organized than the separate amino acids. Nevertheless, peptide links may be driven into existence by coupling with ATP. Each peptide link reduces the entropy of the Universe so much—proteins are such structured, ordered entities—that three ATP molecules must break up to raise each peptide link into existence. But this is no great problem (at least, it has been solved by evolution). The busy activity of enzymes within cells marshalls the necessary ATP to build peptide link after peptide link, following the plan supplied by the template DNA in the nucleus of the cell. Thus you and I emerge.

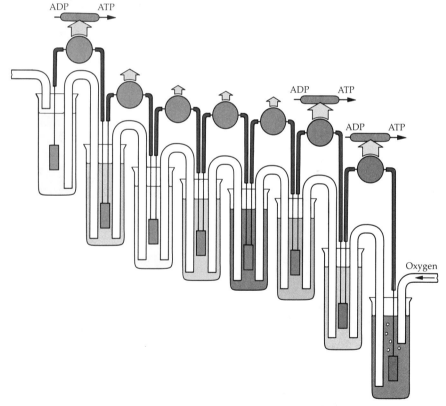

The reactions going on in mitochondria can be compared to a series of electrochemical cells. Each one extracts a little of the energy, and produces work. In three of them, the work produced is enough to attach a phosphate to ADP, and hence to "charge" a molecule of ATP.

9 PATTERNS OF CHAOS

We are now equipped to explore an everyday familiarity, the idea of *structure*. In order to do so, we must first recognize that the discussion of thermodynamics presented so far has glossed over one important area: we have not considered the consequences of a *flow* of material through systems. Ordinary thermodynamics concentrates on *closed* systems, in which matter does not dribble in or out; but a living body is *open*, and matter is ingested as food, drink, and air, and in due course is discarded.

However, our discussion so far has not been useless. In fact, the idea of the dissipation of matter and energy originally stimulated by the classical formulation of the Second Law carries over into this richer class of processes. This is another example of how the child of a concept—here, the idea of dispersal as the underlying rationale of the Second Law—sometimes goes beyond its parent. Now we shall explicitly turn to open systems, and see how they show new, rich phenomena. We shall see how a purposeless flow of energy can wash life and consciousness into the world.

Life is important, not least because it grants us opportunities for enjoyment. We shall therefore take a few extra moments to revel in the pleasure of capturing the apparently dissimilar in a single intellectual net. One of the great aspects of science is its deeper than poetic ability to perceive the underlying unity of things.

Structure

Everyone knows what is meant by a *structure* (see figure on next page). A structure is an arrangement of particles, such as atoms, molecules, or ions. For example, a crystal is a definite structure. It is distinct from a gas, a liquid, or even a splodge of butter, because in these the arrangements of particles are indefinite. Whereas in a crystal we can be sure to find a particle at some definite location relative to another (see figure on page 181), in the "structureless" states of gases, liquids, and amorphous solids, the relative locations of particles are indefinite (see upper figure on page 182).

We can summarize these remarks (and sow the seed for the generalization) by saying that the particles of crystalline solids are arranged *coherently:* the locations of particles are *correlated*. In contrast, in gases (and to a smaller extent in liquids) the locations of the particles are largely incoherent: the locations are uncorrelated. The idea that *structure signifies coherence,* with orderly regiments of particles, whereas *lack of structure signifies incoherence,* with a hodge-podge of locations, neatly captures solids as structures, but allows gases to escape as structureless.

Structures in the world take a variety of forms. Here we see two structures in the everyday sense. We shall shortly extend the range of what is meant.

This preliminary definition of "structure" in terms of a substance composed of coherently arranged particles can be refined in order to pin down the nature of liquids more precisely. When the locations of particles in liquids are measured (with one of the diffraction techniques now so widely used in structural analyses of solids), we find a fairly definite arrangement of particles *locally,* but the further from a given particle we look, the less certain we can be that a particle will be found at a location this local order leads us to expect. In other words, the particles become more jumbled with distance (see lower figure on page 182). This is expressed by saying that solids have *long-range order*. They have a global structure, a large-scale co-

A crystal is a structure consisting of a coherently arranged array of particles. This illustration shows what the inside of a crystal of common salt is like. The green spheres are chloride ions; the red spheres are sodium ions.

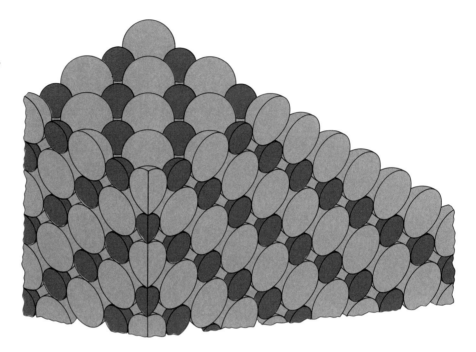

herence, in the sense that the locations of the particles are predictable for great distances (to the edges of crystals, for instance). Gases are almost completely devoid of this global structure. The locations of their particles are incoherent within even very short distances. Liquids are intermediate (as our intuition suggests). They have only *local* structure and lack global structure: within short ranges (relative to a few molecular neighbors) the locations of the particles are coherent, but at large distances they are incoherent. There are subclasses of liquids. For instance, *liquid crystals* have long-range order in some directions, but not in others (see upper figure on page 183). They are therefore liquids in some directions but solids in others (the odd optical properties that result from this anisotropy are the basis of their use in calculator, watch, and TV displays).

Now we shall refine the primitive definition of structure, and enlarge its meaning. *From now on we shall regard structure and coherence as synonymous.* Wherever we observe the onset of coherence, we shall regard it as the emergence of structure. Specifically we shall recognize that coherence need not be merely correlation in *space,* as in ordinary physical structures, but may also be (and this is the crucial point) coherence in *time.*

A gas is not a structure: it consists of a chaotic collection of uncorrelated particles undergoing random motion. There is no coherence in either location or motion.

A liquid is locally a structure, not globally. Although we can be quite definite about the location of nearest neighbors (in the illustration there is a triangular arrangement of many of the neighbors), more distant neighbors are at less predictable locations.

With the generalization stated, we turn to see what the classification captures. In our new net we find old fish, because solids are still caught. But we also find two new fish. One is a type of structure that survives only while energy is being dispersed: these are the *dissipative structures*. They include people. The other is a type of structure that is more abstract, but its existence brings the steam engine full circle; we shall discuss it later.

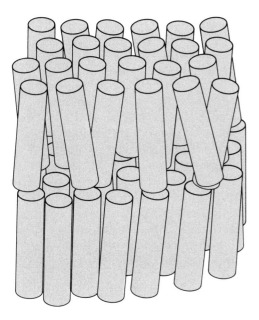

In a liquid crystal, *such as are used in some calculator and watch displays, there is order in some directions but not in others. In this* smectic *state of a liquid crystal, the molecules are arranged in regular planes, but there is disorder within the planes.*

Dissipative Structures

Dissipative structures are structures that arise as a consequence of dispersal. They include some of the fleeting structures of the world, and as soon as the flow of energy or matter ceases, they are lost.

Some of these dust-to-dust structures are biological; others are physical. One of the first to be described was the pattern of cells that forms in a liquid when convection occurs between two horizontal surfaces, the lower surface being hot and the upper cool (see figure below). When the temperature difference between the two plates is low, there is a chaotic distribution of the moving particles of the liquid. However, when the temperature difference is great enough, the *Bénard instability* occurs, and the liquid shows the structure depicted on the next page.

There are two important points to make about this pattern. First, when it occurs, the rate of generation of entropy in the Universe is *increased*, because now energy is being dissipated more rapidly as it flows in an orderly way within the cells and passes from the hot source to the cold sink. Second, along with the more rapid production of entropy there is a *structure* where no structure existed before (or, to be more precise, a global structure has superseded one that was merely local). As soon as the temperature differential is removed, the global structure reverts to local structure, and the convection cells disappear. The structure is sustained by the *flow* of energy, and as soon as that ceases the structure decays.

Dissipative structures are found in chemistry too. In forming these structures, chemical reactions give rise to periodically varying concentrations of substances. The periodic variation may be in time and in space. In time variation, one species gives way to another, and then is regenerated, but only to decay again. In spatial variation, regions of different substances form patterns in the reaction vessel. Such processes underlie the activities of life, and are far from being mere laboratory curiosities. For example, the heartbeat is a periodic process in time, and is sustained by a complex of

The Bénard instability *sets in when a liquid is heated through the lower plate (and cooled at the upper). Two of the convection currents that are established are shown.*

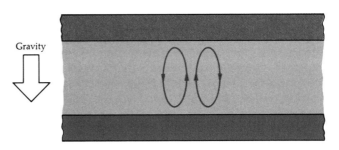

Gravity

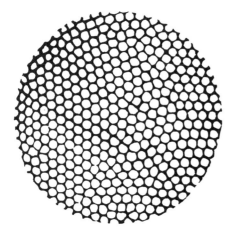

Spatial pattern of convection cells, viewed from above in a liquid heated from below.

chemical reactions that oscillate. Spatially periodic reactions include the organization of cells into bodies, and the patterns on the flanks of zebras and cats closely resemble the striations that arise in some chemical reactions where the components of the reaction are allowed to diffuse through the medium.

There are analogies between the abundances of reactants and products in reaction mixtures and the populations of species of animals in the real world: one reactant in a reaction may be regarded as prey for the other, the predator. The rise and fall of concentrations then become the rise and fall of competing populations. Therefore, we need not deal with remote and unfamiliar chemicals to explore the important concepts here. It is conceptually easier but just as relevant for us to deal with more readily visualizable objects, such as foxes and rabbits, or any other interdependent species of predatory carnivores and herbivores. The argument that allows us to explain the existence of dissipative structures—populations that vary periodically in either space or time—in the countryside (and therefore by analogy also in test tubes) then runs as follows.

Rabbits (R) eat grass (G). Let us suppose that there is a constant and inexhaustible supply of grass. Then the simultaneous presence of grass and rabbits enables the population of rabbits to increase indefinitely. We can symbolize this as:

$$\text{Rabbits} + \text{Grass} \longrightarrow \text{More rabbits.}$$

A chemist might denote the process as $R + G \rightarrow 2R$. The fact that grass is always provided in this land of rabbits (or that a supply of reactant G is always dribbled into the reaction vessel) is akin to the supply of energy in the Bénard problem. Soon the overall process will be seen to be dissipative in much the same way.

The fact that this Rabbit/Grass reaction proceeds spontaneously in the direction of more rabbits is a working out of the consequences of the Second Law. Although the activities of the rabbits may appear purposeful to the superficial observer, at root the reaction is a complexly geared channel down which the Universe is slipping as energy is dispersed. Rabbits are formed from grass (and therefore ultimately are formed by the Sun), but as they take on structure, greater disorder is generated elsewhere. In an actual Rabbit/Grass consumption, the processes involved are extremely complex; nevertheless, the generation of new rabbits out of older rabbits that consume grass (and the free energy that provides) is only a part of the intricate web spun by the collapse of the Universe. The hot spots of the Universe (the Sun, and here the matter composing rabbits and grass) are cooling, and the intricacy of the pathways through it results here and there (at the expense of some grass) in a local abatement that we other (human) abatements can recognize as rabbits. The processes of biology and chemistry, rich and extraordinary as they might seem (and with the scientific eye extinguished and the poetic eye turned on, rich and extraordinary as they are), are no different in principle from cooling.

In the same bucolic scene where the rabbits frolic (below), there lurk foxes (F). Foxes prey on rabbits. Just as grass is consumed by rabbits to produce more rabbits, so rabbits are consumed by foxes to produce more foxes:

$$\text{Foxes} + \text{Rabbits} \longrightarrow \text{More foxes.}$$

The chemist, with a more dispassionate eye, a simpler process, and a simpler notation, might symbolize this by the reaction $F + R \rightarrow 2F$.

A bucolic scene in the Mark I universe. White rabbits lie in the grass, but red foxes lurk. Some (blue) pelts are littered around.

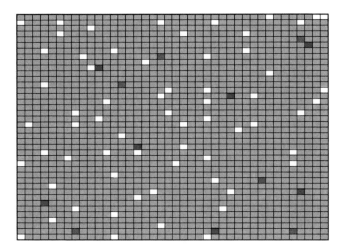

Foxes are subject to predation just as much as rabbits, and, even though they reproduce, they also die and are hunted and killed. Therefore

$$\text{Foxes} \longrightarrow \text{Pelts,}$$

which the chemist might write as $F \rightarrow P$. The pelts, or the chemist's products, are washed out of the ecological system (or the reaction vessel, or the biological cell) and play no further immediate role. However, we can view them as a drain on the energy supplied initially by the grass. The ecological system, like the flask and a cell, possesses a flow.

The crucial steps in the sequence are the ones known technically as *autocatalytic*. An autocatalytic reaction is one in which the products of some step take part in an earlier step. This is like positive feedback: the presence of a substance stimulates the production of more of that substance. In some situations the analogy is with negative feedback: then the presence of a substance inhibits production of more of that substance. In the present scene there are two positive-feedback autocatalytic steps. One is the production of rabbits from rabbits (in effect, by a rabbit consuming grass), because the more rabbits there are, the more rabbits will be generated: an unrestrained autocatalysis would result in a tropical storm of successive generations of rabbits. The second autocatalytic step is the production of foxes from foxes (in effect, by a fox consuming a rabbit). If the supply of rabbits were indefinite, and if the foxes were not too heavily culled, then this step would in due course lead to a storm of foxes.

As a result of the rabbit autocatalysis, there is the potential for a surge of rabbits (there is always plenty of grass in this rabbits' Eden). However, if the fox population grows, there may also be a surge in their numbers as a result of their own autocatalysis. If this surge occurs, a decline in the number of rabbits will ensue, for foxes need rabbits to make more foxes. However, an autocatalysis step magnifies swings down as well as up; if rabbits are consumed, there will be a sharp decline in their production and therefore in their population. But then the fox population will decline as the rabbits come to be in short supply. When the fox population plunges down, the rabbits have time to recover. This they can do quickly, on account of their autocatalysis; but that allows the fox population to recover in a surge too, for the same reason.

It should now be clear that there is a *periodic* oscillation of populations, with a surge of rabbits followed by a surge of foxes, a sharp decline in rabbits, a sharp decline in foxes, a new surge of rabbits, etc. (see figure below). As an alternative to displaying the changing populations of rabbits

The variation of the populations of rabbits and foxes with time. The periodicity corresponds to the emergence of an ecological structure.

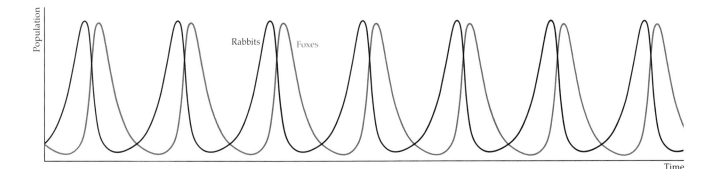

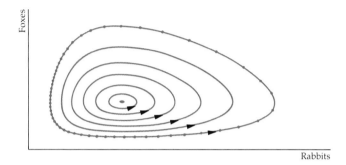

An alternative way of showing the structure is to plot the populations of rabbits and foxes on a single graph, as has been done here. From a given starting population (some point in the plane) the populations evolve, but return periodically to the same value. Each starting point determines a definite trajectory. The point at the center corresponds to a perfectly stable system of rabbits and foxes, with unchanging populations even though birth and death continue.

and foxes by two lines, it is neater to show them on a single graph (as above), in which the horizontal axis represents the rabbit population, the vertical axis the fox population. Time is marked by the dots on the curve. As time goes on, the populations come and go as represented by successive points on the graph, and after a while (which depends on how fast the foxes eat the rabbits, and on how fast they each breed) the cycle begins again. The behavior of the populations with various degrees of fecundity and hunting and avoiding skills can be explored by using the *Population* program listed in Appendix 3.

The point of this discussion may now be made. The periodic cycle of populations is a *structure*. It possesses a coherence in time: the population at any future instant can be predicted from what it happens to be now, and the numbers of species change cyclically. Moreover, the structure is *dissipative*, because the maintenance of the coherence depends on the sustained input of grass and on the culling (by natural or artificial means) of the foxes. There is a flow of energy through the system, the input being represented by the supply of grass (which represents energy captured from the Sun) and the output being represented by the removal of the pelts from the immediate ecosystem into the world of fashion.

Reactions and behavior of the type that we have just described can also give rise to *spatial* patterns. We have seen that the populations of each species comes and goes, but a more detailed model shows that the populations come and go in patterns in the field. In the reaction we have modeled so far, we have made many simplifying assumptions (as you will see from the description of the *Population* program in Appendix 3). For instance, the foxes and rabbits are assumed to be so well-fed that they can hardly move, and eat only their immediate neighbors, neighbors who happen to have been born there (a more realistic ecosystem analogy would replace the rabbits and foxes by competing plants). In order to show spatial patterns, we must allow the rabbits to gambol and the foxes to lurk more actively. That is, the reaction species must be allowed to *diffuse* through their Eden.

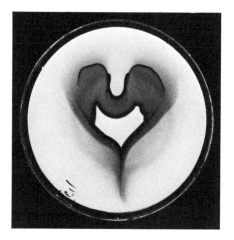
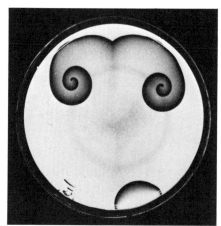
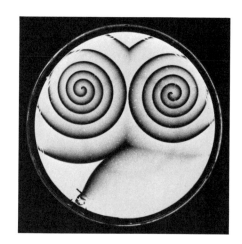

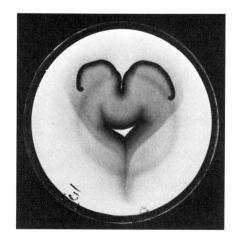

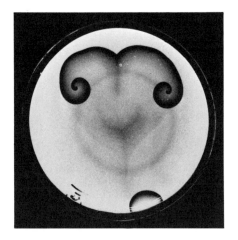  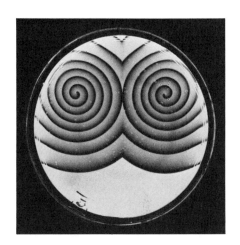

Spatial periodicity is shown by some chemical reactions: this illustration shows a reaction at different times after its initiation.

There are well-documented examples of spatial patterns arising from actual chemical reactions of the same general kind as the ones we have been describing. The sequence of figures here shows what happens in one of them. The emergence of a spatially coherent structure is very plain: the structure itself is very beautiful. Now chaos is generating something close to art.

The Emergence of Complexity

We are beginning to see that individual processes, each one of which may increase the entropy or enhance the chaos of the Universe, can give rise to structures of great complexity. When we observe something with an intricate structure, we therefore should not jump to the conclusion that it is a consequence of design: it may be the outcome of a string of steps, each one purposeless, each one wandering into its destiny as the Universe sinks into chaos. Paley's famous watch summarizes the argument for an active designer: if I found a watch, he argued, then the intricacy of the mechanism would leave me in no doubt that it had been designed, and that there had once, at least, been a designer. He went on to argue that the natural world has an even greater intricacy of construction, and that a traveler coming across the world would be left in no doubt that it too had been Designed, and that there had once, at least, been a Designer. But Paley's argument is false. When we come across a rabbit, there is no need to regard it as de-

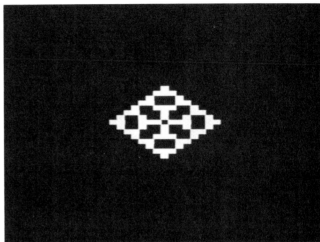

Several stages of the patterns that emerge in the course of playing the game Reproduction *beginning with a single, central seed.*

signed. Rabbits have emerged as a pathway by which the Universe degenerates and the quality of energy degrades. Rabbits, like primroses, pigs, and people, are a part of the great network, the cosmic interconnection that allows temporary structures to emerge as degeneration ineluctably lowers the Universe toward its final equilibrium.

There are plenty of ways of seeing that a network of interdependent simple processes can give rise to complexity, and can therefore perhaps delude the observer into suspecting design. We can explore the point by using two of the mathematical games that were developed for playing with counters on a board and that are quite readily adapted for playing in a lazier way on a computer.

The first game, *Reproduction*, was devised by Stanislav Ulam; I have modified it slightly in order to achieve more striking illustrations. The original game runs as follows. A counter is placed at the center of the board. The next generation is formed by placing a counter on any empty square that has one and only one occupied immediate neighbor. (Neighboring squares are taken to be the four to the north, south, east, and west, not the diagonal neighbors.) Then the rule is applied again, after which the grandparent generation of counters is removed (the original seed at this stage). After the next round of generation of children, the second generation, also now grandparents, is removed. The procedure is then allowed to continue indefinitely.

 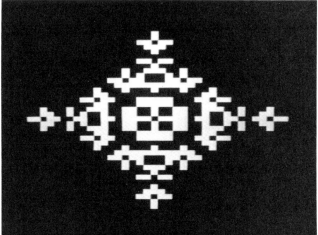

The patterns that sprout from this seed are shown above. The patterns evolve and take on a striking complexity even though the underlying processes are extremely simple. Even more complex patterns may be obtained by placing more than one seed on the board initially. A simple modification to the game, which sharply increases the complexity of the emerging pattern, is to allow the grandparent generation to die as before (so that children can be born in their vacated places), but to mark their passing with a tombstone. If each passing generation is marked with a randomly chosen color, then we obtain some of the striking patterns shown on the next page.

Note how even a tiny increase in underlying complexity has brought about a significant increase in the perceived complexity. In the real world of interdependent processes, where innumerable events are geared together, the perceived complexity may appear to be overwhelming, and might easily delude the observer into thinking that the world is intrinsically complex. The standpoint we adopt, however, is that if simplicity can concatenate and emerge into complexity, then there is no need to assume an awesome underlying complexity. What awe we have ought to be reserved for the richness of the ways in which simplicity can masquerade as complexity.

A second example will illustrate another aspect of simplicity's masquerade. Probably the most famous of all mathematical games is the one

If earlier generations are entombed in a randomly chosen color (the same one for each member of a generation; the undertakers conform to the prevailing fashion), the patterns generated by Reproduction *become more complex.*

devised by the Cambridge mathematician J. H. Conway. This is the game of *Life*. There are only two rules, and both are very simple: they are the rules of birth and death:

Birth: A counter is born in any square with three and only three neighbors;

Death: A counter dies of isolation if it has fewer than two neighbors, and it dies of overcrowding if it has more than three.

(Neighbors in *Life* are taken to be all eight squares around the square of interest; so they include diagonal neighbors as well as those to north, south, east, and west.)

The application of these rules is illustrated below. The game proceeds by a complete pass through the existing generation of counters, marking the locations where births and identifying deaths are about to occur. Then all the newly born counters are added, and the newly dead are removed, in a simultaneous wave. Hence a newly born counter does not beget children until the following generation, when all potential begetters beget together, and the potentially dead do not die until all have been identified and have made their contribution to the begetting.

The behavior that emerges from this set of rules can be illustrated by two primitive examples and a third that has more bite. (Others will be found described in the sources in the bibliography.)

An illustration of the rules of **Life**. *Successive columns denote successive steps. Column 1 shows six seeds. In column 2 the finger of death is laid on in black. In column 3 the locations of births are marked (the fingered still contribute). In column 4 all births and deaths are turned on, and the new generation is established.*

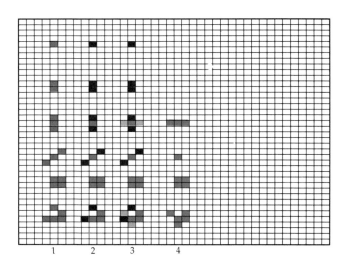

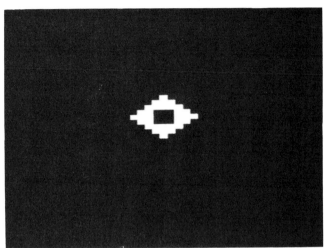

One seed for Life, *two intermediate patterns, and the final, stable pattern that emerges after 16 generations.*

First, the pattern above, left, grows into the flower (after 16 generations). The pattern is more complex when previous generations are entombed with randomly selected colors (see below). We should note two features. First, a species of complexity has emerged from a simple seed with very little intrusion of either design or purpose. The light rein of simplicity has allowed a nascent complexity to flourish. Second, the pat-

If deaths are marked in randomly selected colors (one per generation), a more complex pattern emerges.

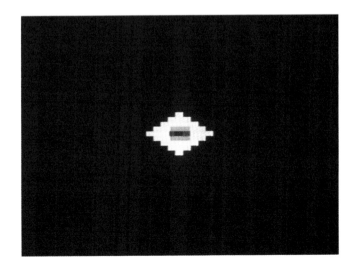

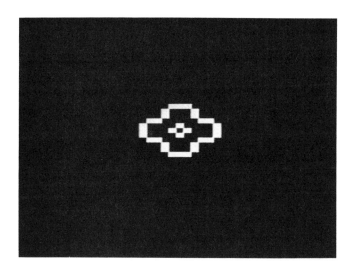

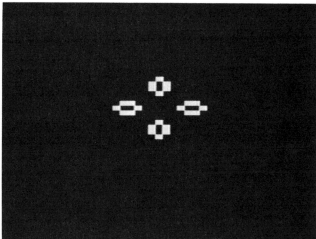

tern is *stable*. Once the rules have brought the seed to the flower, no further change occurs, because all deaths and births are blocked. Not only are the rules midwife to a complex pattern, but they also nurse it into eternity. The rules not only generate a stable structure, but ensure its stability too.

For a second example we take the simple pattern shown at the top of the next page. When the game is run, this little "lemming" wriggles across

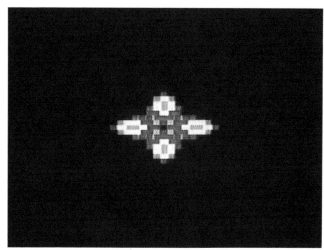

The seed shown here wriggles across the universe in an apparently purposeful journey to its edge. Some call it a glider; we call it a lemming. If the lemming's earlier generations are marked in color, it leaves a vapor trail of tombstones marking its passage.

the page, returning to its initial shape after each fourth generation, and moving apparently purposefully to the edge of the universe. The spoor of its passage can be marked with randomly colored tombstones to give the vapor trail shown above. The apparent purposefulness of the lemming's behavior is an *illusion:* it is just playing out its nature according to the simple physical laws of the little universe it inhabits.

The third example uses the more complicated seed shown on the next page. It is probably impossible to predict what contortions *Life* will take it through without playing out the game. In fact, the seed grows into a pattern which reemerges after thirty generations, but in the process gives birth to a lemming. As the thirty generations of gestation of the next lemming take place, the first lemming wriggles off on its purposeless way to the edge of the universe. The original pattern thus gives birth to lemmings into eternity, and each one scurries inevitably to its doom. Now we have simplicity giving rise to complexity (in the form of patterns), to stability (in the form of periodic regeneration), to purposefulness (in the form of apparent determination to pursue annihilation), and to potency (in the form of an infinite power of generation).

The point of these games is that each one shows that attributes of our present Universe, such as complexity, stability, and apparent purposefulness, emerge as the consequence of simple events played out under a gentle rein of rules. Admittedly, the games are only analogies, and may seem to be no more than allegories. However, in many ways they do model processes in the real world. In particular, when there is a network of inter-

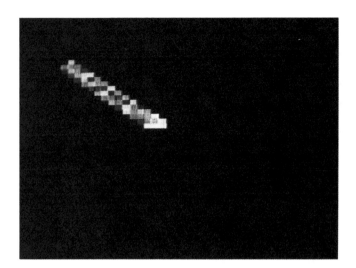

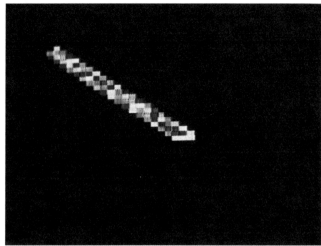

dependent chemical reactions, each of which is a one-way channel through which the Universe can sink into chaos (but cannot return to an earlier state, because the spontaneous reaccumulation of order is so improbable), there may emerge consequences as complex as consciousness. Nothing is more remarkable than consciousness, and nothing is more awesome than that its heart is simplicity.

This complex pattern grows into a cannon which then fires off lemmings forever.

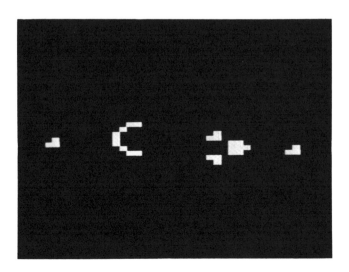

The Apotheosis of the Steam Engine

One of the first lessons that came from the steam engine, when we followed the acute reflections of the classical thermodynamicists, was the distinction between heat and work. We went on to see, as Clausius had recognized, that work and heat are distinct at a microscopic level, in the sense that work is the transfer of energy as coherent motion, heat is its transfer as incoherent motion. In particular, we recognized that work is a method, not a thing, and then proceeded to unfold the implications, especially those based on the distinction as expressed by the Second Law. One consequence of the unfolding of the implications was that we refined the concept of 'structure'.

Now, however, we come to the point where we can recognize that, according to the definition we now adopt, *work is a structure* too! This is true because we have decided to identify coherence with structure, whatever form the coherence might take. If a gas does work by pushing out a piston, then the coherent motion of the atoms is a type of structure.

Moreover, the coherent motion we recognize as work depends on the flow of energy: by a refinement of Carnot's viewpoint, we saw that the flow of energy from hot to cold was the stream that turns an engine, and gives rise to the generation of work. Hence, since the structure we call work disappears when the flow of energy ceases, work is a *dissipative structure*. In this sense, work becomes a thing.

Now we can begin to see the universality of the Second Law in a new light. The complete conversion of heat into work would correspond to the spontaneous emergence of a structure, as the incoherent motion of particles in a system emerged into the outside world as the coherent motion of the particles that constitute pistons and weights. The structure we call work can no more emerge spontaneously than can any other structure of a more conventional kind; all the Second Law is denying is that structures have ever been observed to emerge spontaneously out of disorder. Cathedrals, houses, cows, and people cannot emerge spontaneously; work cannot spontaneously emerge from chaos. The Second Law is a global denial of the emergence of spontaneous structure.

This does not mean that structures cannot ever emerge. We know that we can generate work, and we have seen that conventional structures of all manner of complexity can come into being. The emergence of a plant from a seed, and of a person from a parent, is exactly like the generation of work in an engine. In each situation a coherence is being impressed upon a local part of the Universe, be it the atoms that lie in the vicinity of the roots, the atoms that constitute ingested food, or, most simply of all, the particles that form a piston. In each of these situations too, the coherence is intrinsically transient, and crumbles into incoherence when the structure ceases to be driven by a flow of energy. Death comes to a piston, as to a person, when dissipation ceases. Dust—incoherence—goes to dust; between dusts there is the ramified structure of life. To live we must dissipate and sustain our fleeting disequilibrium, for equilibrium is death.

The coherence of structure emerges if it is driven by the generation of incoherence elsewhere. A local structural coherence can form if it is coupled to a greater crumbling into disorder elsewhere. That elsewhere may be nearby, as when a molecule of ATP crumbles to ADP within a cell. But it may be far away, as when a pair of nuclei fuse in the depths of the Sun, and release the energy of combination in a direction that, in due course, brings it to Earth, where it may be captured by a growing blade of grass. The *simplest* example of construction at the expense of chaos is in an engine, where the incoherence generated in the cold sink exceeds the coherence imposed on the piston. The combined effects of the events taking place in the hot source and the cold sink is the disruption of structure, for energy has lost its quality as a result of its transfer from hot to cold. But the gearing is such that some construction may occur in the course of its dissipation, and the particles of the piston have taken on a structure. That structure is conventionally regarded as work.

In exactly the same way (in principle, that is, not in practice), we dissipative structures live off the disruption of the world elsewhere, and in the process generate a string of structures, until the reactive capacity of our bodies diminishes so much that we can no longer couple effectively to the dissipation around us: then we sink to equilibrium and the grave. But before doing so, we have coupled to chaos to build ourselves, we have used it to enjoy a transient stability, we may sometimes have been deluded into ascribing to our own and others' existence a purpose, and we may have left, to a greater or lesser degree (depending on the opportunities inherited from chance), a lemming's spoor across the history of the world.

We began with the steam engine, and Carnot's determination to improve it. That seed of motivation led to an understanding of the manner in which heat was converted into work, and the recognition that the flow of the stream of energy was its dissipation. Through the eyes of the Second Law we saw that the quality of energy ineluctably declines, and that as it corrupts into chaos, so events became locked into the future. Nature reflects the steam engine, but in a much more elaborate way. Energy everywhere disperses: the world is a globe of corruption. But the dispersion is so channeled and geared together that instead of a single, swift, violent collapse immediately after the world was formed, there is a slow unwinding. In the unwinding, structures emerge locally, and although all are transient, some can last a billion years.

We are the children of chaos, and the deep structure of change is decay. At root, there is only corruption, and the unstemmable tide of chaos. Gone is purpose; all that is left is direction. This is the bleakness we have to accept as we peer deeply and dispassionately into the heart of the Universe.

Yet, when we look around and see beauty, when we look within and experience consciousness, and when we participate in the delights of life, we know in our hearts that the heart of the Universe is richer by far. But that is sentiment, and is not what we should know in our minds. Science and the steam engine have a greater nobility. Together they reveal the awesome grandeur of the simplicity of complexity.

APPENDIX 1 UNITS

The *International System* of units (SI), which is now widely accepted, is based on the kilogram (kg), meter (m), second (s), and ampere (A). These, together with a few others, are the *base units* of the system, and from them the units for all physical quantities may be derived.

Force is expressed in *newtons* (N). A force of 1 newton accelerates a mass of 1 kilogram to a speed of 1 meter/second in 1 second. Formally, $1 \text{ N} = 1 \text{ kg m s}^{-2}$. (A small, 100-g, apple on a tree experiences a gravitational force of about 1 newton; the force you experience when holding this 500-g book is about 5 N.)

Energy is expressed in *joules* (J). An energy of 1 joule is expended in moving an object through 1 meter when it is opposed by a force of 1 newton. Formally, $1 \text{ J} = 1 \text{ N m} = 1 \text{ kg m}^2 \text{ s}^{-2}$. (Lifting this book through 1 meter requires about 5 J of energy; each pulse of the human heart requires about 1 J of energy.) The joule is a small unit in many applications, and it is common to find energies expressed in *kilojoules* (kJ): 1 kJ = 1,000 J. (The energy required to heat 1 quart, 0.946 liter, of water from room temperature to its boiling point is about 18 kJ.)

Power is expressed in *watts* (W). A 1-watt source supplies energy at the rate of 1 joule per second. Formally, $1 \text{ W} = 1 \text{ J s}^{-1}$. (A 100-watt lamp consumes energy at the rate of 100 joules per second; the human body involved in normal activity is rated at about 100 W, a significant proportion being used to drive the brain.) Power is often expressed in *kilowatts* (1 kW = 1,000 W) and *megawatts* (1 MW = 10^6 W). (The Sun radiates energy at the rate of about 70 MW per square meter of its surface; at the equator the Earth receives a mean annual solar energy flux of around 1.4 kW per square meter.)

Older units are still widely encountered. Force is sometimes expressed in dynes, power in horsepower, and energy in ergs and calories. In dietary calculations the Calorie (as distinct from the calorie) looms large. The following table can be used for making conversions.

Unit conversions

Quantity	Unit	SI equivalent
Force	1 dyne	10^{-5} N
Energy	1 erg	10^{-7} J
	1 calorie	4.184 J
	1 Calorie	4.184 kJ
	1 electronvolt	1.602×10^{-19} J
	1 BTU	1.055 kJ
Power	1 horsepower	0.746 kW

Temperature is expressed in *kelvins* (K). The magnitude of the kelvin is the same as the magnitude of a degree Celsius (centigrade), and so the normal melting and boiling points of water differ by 100 kelvins. On the *Kelvin scale* of temperature, the freezing point of water (0 °C, 32 °F) lies at 273.15 K. The following expressions can be used to convert temperatures on the Celsius (t_C), Fahrenheit (t_F), and Rankine (T_R) scales to the Kelvin scale (T):

$$T = [273.15 + (t_C/°C)] \text{ K},$$
$$T = [255.37 + (5/9)(t_F/°F)] \text{ K},$$
$$T = [(5/9) \times (T_R/°R)] \text{ K}.$$

The Rankine scale is like the Kelvin scale, but the size of the degree is the same as on the Fahrenheit scale; so it is a kind of "old-fashioned" absolute scale of temperature. The thermodynamic expressions (such as the Carnot factor) given in the text and the next Appendix work as well for the Rankine scale as for the Kelvin scale; the former is more widely used in engineering, but the latter is virtually universal in physics and chemistry.

APPENDIX 2 FORMALITIES

This Appendix sets out some of the equations of classical thermodynamics. In some of them we give the form for a *perfect gas,* which is one that conforms to the relation $pV = nRT$, p being its pressure, V its volume, T its temperature, and n the amount of substance (the number of moles). The constant R is the *gas constant:* its value is 8.31451 J K^{-1} mol^{-1}. R is a universal constant, independent of the identity of the gas (it also occurs in expressions not related to gases): it and *Boltzmann's constant, k,* are related by $R = kN_A$, N_A being *Avogadro's constant,* 6.022×10^{23} mol^{-1}. The value of k is therefore 1.381×10^{-23} mol^{-1}.

Thermodynamics

The *internal energy* (U) of a system changes by an amount ΔU when heat is supplied and work is done. If the *heat supplied to the system* is denoted q and the *work done on the system* is denoted w, then $\Delta U = q + w$. For a gas expanding against a pressure p, the work done for an infinitesimal change of volume dV is $dw = -pdV$. If the pressure of the gas is almost exactly balanced by the external pressure throughout the expansion, the process is *reversible.* (In general, a reversible process is one that can be reversed by an infinitesimal modification of the conditions.)

The *enthalpy* (H) of a system is related to its internal energy by $H = U + pV$. The old-fashioned name for enthalpy is "heat content." It is important because knowing the change in enthalpy of a system lets us predict the quantity of energy that it can produce as heat when the process occurs at constant pressure (the common condition for many combustions and other reactions). For changes under constant pressure, $\Delta H = q$. That is, if a reaction corresponds to an enthalpy change of -100 kJ, then under conditions of constant pressure it can liberate 100 kJ of heat. *Exothermic reactions* are accompanied by a reduction of enthalpy (and so liberate heat); *endothermic reactions* are accompanied by an increase of enthalpy (and so draw in heat). The "latent heats" of vaporization and melting are in fact the enthalpy changes accompanying these processes.

Some enthalpy changes[a]

Process	Enthalpy change/kJ mol^{-1}
Ice melting at 0 °C	6.01
Water boiling at 100 °C	40.66
Methane burning	−890
Benzene burning	−3268
Glucose burning	−2816
Water forming from hydrogen and oxygen	−285.8

[a]Temperatures are 25 °C except as specified. One mole is an Avogadro's number of molecules.

Although *absolute* values of enthalpies and internal energies cannot be measured, the *change* in their values can be measured quite simply by observing the transfer of heat during a process. The *enthalpy of formation* of a compound is the enthalpy change that accompanies the formation of a compound from its elements; the *enthalpy of combustion* is the enthalpy change that accompanies the complete combustion (to carbon dioxide and water, typically) of the compound. Some representative values are given in the table above. They may be combined in various ways to give the enthalpy change that accompanies other reactions (for instance, the enthalpy of combustion of methane can be expressed in terms of the enthalpies of formation of methane, oxygen, carbon dioxide, and water).

The *heat capacity* of a system is related to the rise in temperature it shows for a given transfer of heat: the larger the heat capacity, the smaller the rise. The heat capacity depends on the conditions, and it is common to encounter two varieties: the heat capacity at constant volume (C_V), and the heat capacity at constant pressure (C_p). Formally, $C_V = (\partial U/\partial T)_V$ and $C_p = (\partial H/\partial T)_p$. They are related (for a perfect gas) by $C_p - C_V = nR$.

The *entropy* (S) of a system is defined in such a way that when an infinitesimal quantity of energy dq is transferred reversibly as heat, and the temperature is T, then it changes by an amount $dS = dq/T$. It follows from this and the preceding information that the entropy of a substance at a temperature T, $S(T)$, is related to its entropy at $T = 0$ by

$$S(T) = S(0) + \int_0^T \{C_p(T)/T\}dT.$$

Consequently, by measuring the heat capacity from very low temperatures up to the temperature of interest, we can calculate the entropy. *Third-law entropies* are obtained by setting $S(0) = 0$ for perfect crystalline forms of the

The entropies of some substances[a]

Substance	Entropy/J K^{-1} mol^{-1}
Diamond	2.4
Copper metal	33.1
Copper sulfate	300.4
Water	69.9
Ethanol (alcohol)	160.7
Hydrogen gas	130.6
Oxygen gas	205.0

[a]Values refer to 25 °C and atmospheric pressure.

substances. Some values are given in the table above. The entropy change during the course of a reaction can be calculated by adding and subtracting the entropies of the participants in the reaction, just as enthalpies of formation may be combined to give enthalpy changes.

Entropy changes for some simple processes can be calculated. The entropy change during the isothermal expansion of a perfect gas from a volume V_1 to a volume V_2 is given by

$$\Delta S = nR \log (V_2/V_1).$$

The entropy change when an amount n_1 of one perfect gas mixes with an amount n_2 of another under conditions of constant pressure and temperature is

$$\Delta S = nR\{x_1 \log x_1 + x_2 \log x_2\},$$

n being the sum $n_1 + n_2$, and x_1 and x_2 being the ratios n_1/n and n_2/n, respectively. This was the expression used in the discussion of the role of mixing in the text.

The *Carnot factor* for the efficiency of conversion of heat into work was derived in the text: $w/q = [1 - (T_{COLD}/T_{HOT})]$. The corresponding factor for the efficiency of refrigeration is the same, but with the two temperatures interchanged. These expressions may be stated differently if we suppose that the working substances are perfect gases. Then, for a *Carnot cycle*, the efficiency is

$$w/q = 1 - (V_A/V_D)^{\gamma-1},$$

γ being the ratio C_p/C_V (for air its value is 1.40). For the *air-standard Otto cycle* the efficiency is

$$w/q = 1 - (1/r)^{\gamma-1},$$

with r the *compression ratio*, V_F/V_C. For the *air-standard Diesel cycle* the efficiency is

$$w/q = 1 - (A/\gamma Br^{\gamma-1}),$$

with $A = [(V_D/V_C)^{\gamma-1}] - 1$ and B the same except that $\gamma - 1$ is replaced by 1. These expressions enable us to explore how the efficiencies of the engines depend on the dimensions of the stroke and (through γ) the properties of the working gas.

The *Helmholtz free energy* or *Helmholtz function, A*, is defined as $A = U - TS$. The closely related *Gibbs free energy* or *Gibbs function, G*, is defined as $G = H - TS$. The change in the Helmholtz free energy for a process is equal to the maximum work that process can perform. The change in the Gibbs free energy for a process is equal to the maximum work other than expansion work that process can perform under conditions of constant pressure. Thus, for an infinitesimal change at constant temperature, $dG = dH - TdS$; but $dH = dU + pdV + Vdp$ and $dU = dq + dw$. Under conditions of reversibility, $dq = TdS$ and $dw = -pdV + dw_{extra}$. Therefore, putting these all together results in $dG = Vdp + dw_{extra}$, since everything else cancels. If the pressure is held constant, $dp = 0$, and so $dG = dw_{extra}$. Work done *by* the system is the negative of the work done *on* the system; so the maximum work that a system can do when it undergoes a specified change under conditions of constant temperature and pressure is $dw_{extra,by} = -dG$. This is the form used in the text (as $w = -\Delta G$).

Tables of Gibbs free energies have been compiled, and a representative selection is given in the table below. Just as with entropies and enthalpies, we can combine these to obtain the change of Gibbs free energy that accompanies a particular reaction of interest. A major application in chemistry is to predict the *equilibrium constant, K*, of a reaction: if the Gibbs free energy change (under certain specified conditions, those used for the val-

Some changes in Gibbs free energy[a]

Process	Free energy change/kJ mol^{-1}
Formation of water	-237.2
Formation of ammonia	-16.5
Formation of NO_2	$+51.3$
Formation of N_2O_4	$+97.8$
Combustion of methane	-818.0

[a]All processes are at 25 °C and atmospheric pressure. In the cases listed, the formation reaction is the process of forming the compound from the normal gaseous forms of its elements.

ues in the table) is ΔG, then the equilibrium constant may be obtained from

$$\Delta G = -RT \log K.$$

This is a powerful link between thermometric measurements and practical applications to chemistry.

Temperature

A formal thermodynamic connection between the internal energy, the entropy, and the temperature is $T = (\partial U/\partial S)_V$. For a two-level system in which the energy separation of the two levels is ϵ, a consistent definition of temperature is

$$T = (\epsilon/k)/[\log (N_{OFF}/N_{ON})],$$

where N_{OFF} is the number of particles in the lower energy level and N_{ON} is the number in the upper. This shows the explicit definition of the temperature in terms of the difference between the levels, as was mentioned in the text.

The internal energy of the system is given by the expression

$$U(T) = U(0) + \epsilon N_{ON},$$

and the entropy by

$$S(T) = k \log W, \qquad W = N!/N_{ON}! \, N_{OFF}!.$$

N is the total number of atoms, and ! denotes factorial ($N! = N \times (N - 1) \times (N - 2) \times \ldots 2 \times 1$). When the numbers involved are large, it is possible to use the approximate relation

$$S(T) = [U(T) - U(0)]/T + Nk \log q,$$

with $q = 1 + N_{ON}/N$.

SOME SOURCES FOR FURTHER READING

A wide variety of excellent texts cover the ground spanned here, but almost all require more mathematical sophistication than I have asked. The closest in spirit to this exposition, naturally enough, is my own *Atoms, Electrons, and Change* in the Scientific American Library (W. H. Freeman and Co., 1991), which looks at the chemical consequences of the Second Law more closely. I have also explored the philosophical implications of the Second Law in a light-hearted but deeply felt manner in *Creation Revisited* (W. H. Freeman and Co., 1993). As to other non-mathematical introductions, two excellent accounts are *Engines, Energy, and Entropy* by J. B. Fenn (W. H. Freeman and Co., 1982) and *The Refrigerator and the Universe* by M. Goldstein and I. F. Goldstein (Harvard University Press, 1993). My *Elements of Physical Chemistry* (Oxford University Press and W. H. Freeman and Co., 1992) introduce the concepts of thermodynamics at a precalculus level, and my *Physical Chemistry,* fifth edition (Oxford University Press and W. H. Freeman and Co., 1994), deals with the chemical aspects in detail. For excellent and thorough accounts of physical aspects of thermodynamics, I recommend *Heat and Thermodynamics,* sixth edition, by M. W. Zemansky and R. H. Dittman (McGraw-Hill, 1981), a book that has greatly influenced my thinking. Good introductory accounts will also be found in *Thermodynamics* by K. Wark (McGraw-Hill, 1988). A book with deep insights yet straightforward exposition is the classic *Thermodynamics* by G. N. Lewis and M. Randall in the version revised by K. S. Pitzer and L. Brewer (McGraw-Hill, 1961). Applications to engineering problems are treated in the book by Wark mentioned above and in *Applications of Thermodynamics,* second edition, by B. D. Wood (Addison-Wesley, 1982), which gives a thorough account of all the cycles I have mentioned. A useful reference is also *The Principles of Refrigeration* by W. B. Gosney (Cambridge University Press, 1982).

For the thoughts behind thermodynamics, I particularly like *Energy, Force, and Matter* by P. M. Harman (Cambridge University Press, 1982): its subtitle, *The Conceptual Development of Nineteenth-Century Physics* reveals its context. A reflective essay is *The Nature of Thermodynamics* by P. W. Bridgman (Harper, 1961), and a similar thoughtful but slightly more mathematical exposition of the foundations of the subject is *Entropy in Relation to*

Incomplete Knowledge by K. G. Denbigh and J. S. Denbigh (Cambridge University Press, 1985).

The statistical side of thermodynamics is difficult to penetrate without a fairly stiff helping of mathematics, but *The Second Law* (no relation!) by H. A. Bent (Oxford University Press, 1965) is a lively account at an elementary level. Introductory accounts will be found in a variety of texts, such as *Entropy and Energy Levels,* second edition, by R. P. H. Gasser and W. G. Richards (Oxford University Press, 1986), *The Theory of Thermodynamics* by J. R. Waldram (Cambridge University Press, 1985), *Introduction to Statistical Mechanics* by D. Chandler (Oxford University Press, 1987), and *Statistical Thermodynamics and Kinetic Theory* by C. Hecht (W. H. Freeman and Co., 1990), roughly in that order. Thorough expositions of the subject are given in *Foundations of Statistical Mechanics* by W. T. Grandy (D. Reidel, 1988) and *Statistical Mechanics* by A. Münster (Springer, 1974).

The applications of thermodynamics to biochemical processes can be explored through *The Vital Force: A Study of Bioenergetics* by F. M. Harold (W. H. Freeman and Co., 1986) and *Biothermodynamics* by J. T. Edsall and H. Guttfreund (Wiley, 1983). The ever-popular *Biochemistry,* fourth edition, by L. Stryer (W. H. Freeman and Co., in press) should be consulted for the biological context of the concepts mentioned here as well as for consummate visual presentation of material. The hydrophobic interaction and other structural features are treated well in *Biophysical Chemistry* by C. R. Cantor and P. R. Schimmel (W. H. Freeman and Co., 1980), especially in Part I.

Dissipative structures and far-from-equilibrium phenomena are introduced in a variety of accessible texts. Two of the more non-mathematical introductions will be found in *From Being to Becoming* by I. Prigogine (W. H. Freeman and Co., 1980) and *Order out of Chaos* by I. Prigogine and I. Stengers (Heinemann, 1984). The mathematics of such structures is developed in *Oscillations, Waves, and Chaos in Chemical Kinetics* (Oxford University Press, 1994), *Synergetics* by H. Haken (Springer, 1977), and *Chemical Oscillations and Instabilities* by P. Gray and S. K. Scott (Oxford University Press, 1990). The book on *The Timing of Biological Clocks* by A. Winfree in the Scientific American Library (1987) should also be consulted.

An excellent account of mathematical games and the emergence of complex patterns will be found in *The Laws of the Game* by M. Eigen and R. Winkler (Allen Lane, 1982), and in this context I warmly recommend *The Recursive Universe* by W. Poundstone (Oxford University Press, 1987) for its discussion of cellular automata and their real-life analogies. The last three chapters of *Wheels, Life, and Other Mathematical Amusements* by M. Gardiner (W. H. Freeman and Co., 1983), are devoted to the *Life* game. An excellent description of the emergence of patterns on animal pelts is given in *Mathematical Biology* by J. D. Murray (Springer, 1989).

SOURCES OF ILLUSTRATIONS

Chapter opening paintings
George Kelvin

Line drawings
Gabor Kiss

page 1
Deutsches Museum

page 2
Reproduced by permission of The Reference Library Archives Department, Birmingham Public Libraries, England

page 3
Deutsches Museum

page 4
Ann Ronan Picture Library

page 5 (top)
Deutsches Museum

page 5 (bottom)
Deutsches Museum

page 6
Randall M. Fenstra and Joseph A. Stroscio, IBM T. J. Watson Research Center

page 7
Julie Newdoll, Computer Graphics Laboratory, UCSF, © Regents, University of California

page 11
Chip Clark

page 12
United Technologies, Pratt & Whitney Aircraft

page 16 (top)
The Granger Collection

page 16 (bottom)
Archives de l'Académie des Sciences de Paris

page 52
Travis Amos

page 65
Dieter Flamm

page 87
Kim Steele

page 93
Adapted from *Applications of Thermodynamics* by B. D. Wood, Addison-Wesley, 1982

page 128
From *Powers of Ten* by Philip Morrison and Phylis Morrison and The Office of Charles and Ray Eames, Scientific American Books, copyright 1982

page 129
Peter Kresan

page 136
Adapted from *Heat & Thermodynamics* by M. W. Zemansky, McGraw-Hill, 1968

page 141
AT&T, Bell Laboratories

page 151 (bottom)
Lawrence Berkeley Laboratory, University of California

page 163
Irving Geis, from *Biophysical Chemistry Part I, The Conformation of Biological Macromolecules* by Charles R. Cantor and Paul R. Schimmel, W. H. Freeman and Company

page 167
Bettmann Archives

page 180 (left)
Kim Steele

page 180 (right)
Steven Smale

page 184
From *Being To Becoming* by Ilya Prigogine, W. H. Freeman and Company, copyright 1980

pages 188, 189
From *Being To Becoming* by Ilya Prigogine, W. H. Freeman and Company, copyright 1980

page 197 (bottom)
From *Wheels, Life, and Other Mathematical Amusements* by Martin Gardner, W. H. Freeman and Company, 1983

page 200
William Garnett

INDEX

A, Helmholtz free energy, 208
abatement of chaos,
  and construction, 86
  local, 105
  introduced, 41
absolute zero,
  attainability, 41
activation energy, 122
adenosine diphosphate, ADP, 174
adenosine triphosphate, ATP, 173
  formation in mitochondria 176
adiabat, for gases, 17
adiabatic compression, and temperature
    rise, 17
adiabatic cooling, 137
adiabatic demagnetization, 146
    nuclear, 147
adiabatic expansion, for refrigeration, 137
adiabatic process, defined, 16
ADP, 173
air conditioning, 136
air-standard
    Otto cycle, 205
    Diesel cycle, 206
air-standard cycles, introduced, 94
alcohol, structure, 159
alpha helix, introduced, 161
amino acid, 160
amino group, 159
amorphous solids, 179
Arrhenius rate law, 122, 123
Arrhenius, Svante, 122
asymmetry of response,
  reciprocating, 83
  rotational motion, 103
Atkins cycle, as a useless process, 18

atoms
  images, 6
  introduced, 45
ATP, 174
autocatalysis, and dissipative structures,
    186
automobile engine
  Diesel engine, 97, 206
  efficiency, 40
  Otto engine, 94, 205
  turbo-supercharged, 101
Avogadro's constant, 205, 206
Avogadro's number, introduced, 46

Bardeen, Cooper, and Schreiffer, 140
Beau de Rochas, 94
beauty, 200
behavior, and the minimum of rules, 52
Bénard instability, 183
Big Bang
  formation of elements, 13
  microwave background, 142
biosynthesis, and free energy, 172
blood, color, 108
boiling an egg, 164
Boltzmann distribution, 122
Boltzmann probability, 122
Boltzmann's constant, 65, 203
Boltzmann's Demon
  introduced, 67
  and UP-ness and DOWN-ness, 143
Boltzmann's tomb, bridge between macro
    and micro worlds, 81
Boltzmann, Ludwig
  insight into atomic basis, 5
  tombstone, 65

bond length, 109
bouncing ball, example of irreversibility, 59
Boyle, Robert, contribution to gas theory, 16
Brayton cycle
  closed, 102
  indicator diagram, 103
  introduced, 102
  open, 104
BTU, 204

cage structures in water, 164
caloric, as something transferred, 23
  Carnot's acceptance of the idea, 2
calorie, 204
Carnot, *Réflexions*, 2
Carnot cycle
  details, 17
  game, 210
  indicator diagrams at different temperatures, 42
  introduced, 15
  reverse for refrigeration, 130
Carnot efficiency, for refrigeration, 134
Carnot efficiency factor, 40, 207
Carnot engine
  illustrated, 15
  in Mark II universe, 82
Carnot factor, 40, 207
Carnot, Sadi, 1
carrot and cart, 121
cell
  biological, 175
  electrochemical, 174
  nucleus, 175
change, deep structure of, 200
chaos
  and internal combustion engines, 97
  local abatements, 105, 157
chemical reaction, reason for change, 115
chemical bond, described, 108
chemical reactions
  and autocatalysis, 186
  and electrochemical cells, 175
  and energy dispersal, 112
  and irreversibility, 112
  and spatial patterns, 183
  as energy processors, 170
  driving power, 167
  driving power and free energy, 172

entropy changes, 113, 168
  introduced, 108
  model, 111
  oscillating, 183
  spatial periodicity, 189
  work production, 167
Clapeyron, Emile, 15
Clausius statement
  of Second Law, 25
  equivalence to Kelvin statement, 27
  in terms of entropy principle, 33
Clausius, Rudolph, 4
coal, as fuel, 12
coherence
  and structure, 180
  in time, 181
coherent light, 151
coherent motion, degradation into incoherent motion, 61
cold, world record, 147
cold sink, essential feature of engine, 20
cold substances
  at 30 K, 139
  at 3 K, 140
cold worlds, 129
collapse into incoherence, 76
complex numbers, 127
complexity, as interaction of simplicities, 190
compression ratio, and efficiency, 207
compressional heating, as game of ping-pong, 85
consciousness, 199
consciousness as cooling, 107
conservation
  of heat, 2
  of energy, 8, 46
constructive chaos, 157
convection, 183
conversion of heat into work, at microscopic level, 83
Conway, John, 193
cooling
  and Boltzmann's Demon, 66
  as heating, 114
  entropy changes in Mark I universe, 68
  example of energy dispersal, 53
coordinated motion, introduced, 48
corruption, defined and refined, 23

cosmic background radiation, 141
crystalline solids, 180
Cullen, William, 142
cycle
  Atkins, 18
  Brayton, 102
  Carnot, 17

death, for pistons and people, 199
decane molecule, illustrated, 109
decoupling of matter and radiation, 142
degradation of energy, 60
degradation of quality, and life, 172
demagnetization, adiabatic, 143
Demons
  Boltzmann's, 67
  Maxwell's, 67
destruction as construction, 86
diatomic molecule, illustrated, 109
Diesel engine
  efficiency, 206
  introduced, 97
  indicator diagram, 97
  Mark II model, 98
  turbo-supercharged, 101
  two-stroke, 100
discarded energy, and efficiency of engines, 39
dispersal of energy
  and chemical reaction, 112
  interpretation, 62
  introduced, 53
  oil dissolving, 158
  overall importance, 57
dispersal of particles, 76
  mixing, 120
  oil droplets, 157
  summarized, 81
dissipative structure, Bénard instability, 183
  introduced, 182
  Rabbits and Foxes game, 217
  entropy production, 183
  work, 198
dissociation reaction, 114
  role of mixing, 119
DNA
  as template, 165
  molecule, 7
DOWN-ness, 143

dynamic equilibrium, introduced, 72
dyne, 203

e, exponential, 68
eating, 166
  and entropy, 172
ecological system, dissipative structure,
  185
efficiency
  automobile engine, 40
  Carnot calculation, 39
  Carnot cycle, 205
  Carnot's original vision, 1
  defined, 40
  Diesel cycle, 206
  effect of compression ratio, 206
  Otto cycle, 205
  power plants, 40
  upper limit, 41
electric current, generation by chemical
  reaction, 176
electrochemical cell
  introduced, 174
  operation, 175
electron spin, magnetic moment, 143
electronvolt, 204
emergence of order, 124
endothermic reaction, and enthalpy, 205
energy
  capacity to do work, 8
  conservation of, 8
  conservation of total energy, 46
  crisis, 39
  dispersal as a natural process, 53
  distribution and irreversibility, 9
  kinetic and potential, 46
  mill, 20
  primacy in Kelvin's view, 9
  quality, 38
  total in universe, 9
energy dispersal
  and entropy increase, 70
  and chemical reaction, 112
energy magnifier, 135
energy principle, 32
energy processors, 170
engine, as cyclic device, 14
enthalpy, defined, 205
enthalpy changes, table, 204
enthalpy of combustion, 204

entropy
  and quality of stored energy, 38
  crisis, 39
  definition, 34
  first introduced, 30
  increase on gas expansion, 78
  sense of familiarity required, 35
  table of values, 205
  temperature dependence, 206
entropy change
  in course of chemical reaction, 113
  in course of reaction, 168
  in surroundings, 114
  in system, 114
  isothermal expansion, 207
  mixing, 207
entropy meter, 35
entropy principle, 32
entropy production, and dissipative
  structures, 183
equilibrium, and probability, 73
equilibrium constant, and free energy, 207
erg, 204
escape velocity, 138
essential amino acids, 159
ethanol, 159
evaporation and cooling, 142
exothermic reaction
  and enthalpy, 203
  introduced, 167

Fermi sea, 109
final state of universe, in Mark I version,
  54
First Law
  introduction of energy, 29
  summarized, 8
fluctuations
  and departure from equilibrium, 73
  introduced, 47
  of temperature, 57
food, and entropy, 172
force
  displaced from primacy, 8
  units, 201
formation reactions, free energy, 206
fossil fuel, mining in time, 12
four-stroke engines, 100
free energy, and equilibrium constant, 207
  and spontaneous change, 171

defined, 169
  Gibbs, 206
  Helmholtz, 206
  in electrochemical cell, 175
  introduced, 165
  table of values, 206
fuels, example of chemical change, 107
functional thermodynamics, 10
fusion devices, magnetic field generation,
  140

G, Gibbs free energy, 206
gadolinium sulfate, used for cooling, 143
gas
  chaotic structure, 76, 180
  cooling on expansion, 138
  dispersal of particles, 77
  entropy change on expansion, 78
  perfect, 205
  properties, 16
gas constant, 203
Gibbs free energy
  introduced, 169
  defined, 206
Gibbs, Josiah Willard, 169
glucose, as fuel, 173
glycolysis, 174

H, enthalpy, 203
heartbeat, 183
heat
  as method of transfer of energy, 23
  as stimulation with incoherent motion, 48
  caloric, 2
  conservation of, 2
  flow from hot to cold, 25
  incomplete conversion into work, 24
  mechanical equivalence, 3
  production by primitive methods, 11
  taxed when converted, 21
heat and work, distinction at atomic level,
  48
heat capacity
  definition, 204
  in entropy measurements, 36
heat engine, as energy mill, 20
heat flow, as shorthand, 24
heat pump, 135
helium, *see* liquid helium

Helmholtz free energy
  defined, 206
  introduced, 169
hemoglobin, 161
  as oxygen transporter, 176
  structure, 163
horsepower, 201
hot matter, 147
Hume, David, reports of miracles, 75
hydrocarbon, in water, 159
hydrocarbon molecule, illustrated, 109
hydrogen bond, basic structure, 160
hydrogen molecule, structure, 109
hydrophilic, 164
hydrophobic effect, introduced, 159

imaginary time, 128
incoherent motion, introduced, 48
indicator diagram
  Brayton cycle, 103
  Carnot cycle, 15
  Diesel engine, 97
  introduced, 15
  Otto cycle, 94
  Stirling engine, 90
infinite temperature, 150
internal combustion engines, introduced, 94
internal energy, 203
International System of units (SI), 201
interplanetary missions, and cooling, 137
ion, in metals, 109
ions, introduced, 45
iron burning, 111
iron oxide, schematic illustration, 110
irreversibility
  and probability, 74
  bouncing ball, 59
  related to vastness, 55
isolated system, defined, 30
isotherm, for gases, 16
isothermal magnetization, 146
isothermal process, defined, 16

jet engine
  as complex device, 12
  Brayton cycle, 104
  illustrated, 12
joule, J, unit, 201
Joule, James, 3
Joule-Thomson effect, 138

k, Boltzmann's constant, 65
K, Kelvin temperature, 204
Kelvin, temperature scale, 40
kelvin, K, unit, 202
Kelvin statement
  at negative temperature, 152
  equivalence to Clausius statement, 26
  of Second Law, 24
  relation to entropy principle, 32
Kelvin, Lord, meeting with Joule, 4
kilojoule, kJ, unit, 201
kilowatt, kW, unit, 201
kinetic energy, introduced, 46
Krebs cycle, 176

lasers, 151
laws of thermodynamics, sardonic version,
  42
lemming, life game, 195
life
  and open systems, 179
  game, 193
liquid air manufacture, 139
liquid crystals, 181
liquid helium
  production, 140
  superfluidity, 140
local order, 180
localization of energy, 113
long-range order, 180
low temperatures, 129

magnetism, 143
magnetization, isothermal, 143
Mark I universe, introduced, 50
Mark II universe, introduced, 51
Mark III universe, introduced, 51
Martini, 159
mathematics, role of, 10
maximum work, and free energy, 208
Maxwell's Demon, 67
mechanical equivalence of heat, 3
megawatt, MW, unit, 201
metal
  electrical conductivity, 110
  electronic structure, 109
  malleability, 110
microwave background radiation, 141
miracles, 75

mitochondrion, 174
  as electrochemical cell, 176
mixing
  as dispersal, 117
  role in reactions, 117
mole, 206
molecules
  examples of structures, 109
  introduced, 45

natural processes, see spontaneous
    processes
nature's dissymmetry
  and nature's tax, 21
  differences between heat and work, 13
  Kelvin and Clausius formulations, 24, 25
  ordered and disordered motion, 13
  recognition, 14
negative temperature
  heat engines, 152
  introduced, 149
new temperature, 151
newton, N, unit, 201
nitrogen dioxide dimerization, 114
  contributions to entropy, 116
nuclear adiabatic demagnetization, 147
nuclear spin, negative temperature, 153
nucleon, 148

oil
  as fuel, 12
  dispersal of, 157
ON*-ness, introduced, 51
ON-ness, introduced, 50
open systems, 179
ordered motion, extraction of, 13
organelle, 174
oscillating reactions, 183
Otto cycle, introduced, 94
Otto engine, Mark II model, 97
oxygen molecule, illustrated, 109

Paley's watch, 189
paramagnetic material, 143
particles, introduced, 45
pendulum motion, and energy
    conservation, 46
Penzias and Wilson, microwave
    background, 141
people, as dissipative structures, 182

peptide chain, 161
peptide link, 161
  formation with ATP, 177
phosphate groups, 173
picnic, 129
plasma, 140, 148
pointer readings, false sense of familiarity, 35
population, game, 187
population oscillation, rabbits and foxes, 186
populations, at negative temperatures, 150
positive feedback, and autocatalysis, 186
potential energy, introduced, 46
power, units, 201
Powers of Ten, 127
primary structure of protein, 161
probability
  and construction, 86
  and equilibrium, 75
  of states of the universe, 73
protein
  primary structure, 161
  quaternary structure, 164
  synthesis, 165
  tertiary structure, 164
proteins introduced, 159
purpose
  as illusion, 196
  lack of, 58
pyruvate ions, 174

q, heat, 205
quality
  and entropy, 38
  decline of, 38
  of energy, 38
quasistatic process, defined, 34
quaternary structure of protein, 164

R, gas constant, 203
rabbits and foxes, 184
  game, 217
radiation/matter decoupling, 142
radiation/matter recoupling, 148
Rankine scale, 202
rate of process, introduced, 107
rate of reaction, role of chaos, 121
refrigeration,
  adiabatic demagnetization, 146

adiabatic expansion, 137
  Carnot efficiency factor, 134
  evaporation mechanism, 142
  Joule-Thomson effect, 138
  power requirements, 134
  thermodynamics, 130
  work required, 134
regenerator, in Stirling engine, 88
reproduction, game, 190
respiration
  as rusting, 108
  terminal respiratory chain, 176
Rogue, Jack and Jill, 26
rules, none versus imposed, 53
rust, iron oxide, 110
rusting, entropy change, 113
  thermodynamics, 108

Second Law
  and cold sink, 21
  and negative temperatures, 154
  Clausius statement, 25
  in terms of entropy, 32
  introduction of entropy, 30
  Kelvin statement, 24
  summarized, 9
secondary structure of proteins, alpha helix, 161
SI units, 203
sneezing, 86
solar energy, exploitation, 12
space, temperature of, 141
spatial coherence, and chemical reactions, 183
spatial patterns, 187
specific heat, in entropy measurements, 36
spin
  electron, 143
  nuclear, 147
spontaneous change
  chemical reactions, 171
  defined, 31
spontaneous processes, in terms of energy dispersal, 62
star explosions, formation of elements, 13
statistical thermodynamics, 10
steady state, introduced, 55
steam engine
  apotheosis, 198
  as model of nature, 199

as water mill, 2
  efficiency, 1
  universal motor, 1
Stirling engine
  efficiency, 92
  indicator diagram, 90
  principle, 88
  real indicator diagram, 93
  real version, 93
Stirling, Robert, 87
structure
  and coherence, 180
  emergence of, 157
  general definition, 181
  relation to mechanical coherence, 105
  spontaneous emergence, 198
  the concept generalized, 179
Sun
  our debt to, 10, 172
  radiative energy, 203
superconductivity, 140
superfluidity, 140
sustaining cold, 134
system, defined, 31

tax, on conversion of heat into work, 21
temperature
  as imaginary time, 128
  false sense of familiarity, 35
  fluctuation, 57
  in Mark I universe, 55
  infinite, 150
  Kelvin scale, 204
  negative, 149
  new definition, 151
  of space, 141
  Rankine scale, 202
  relation to incoherent motion, 55
  relation to populations, 56
  Zeroth Law, 8
terminal respiratory chain, 176
tertiary structure of protein, 164
thermal equilibrium, and Zeroth Law, 29
thermal motion, as incoherent motion, 48
thermal reservoir, introduced, 16
thermodynamic observer, introduced, 54
thermodynamics
  distinction from dynamics, 48
  functional, 10
  origin of name, 4

thermodynamics, *continued*
  statistical, 10
  survey of laws, 8
thermodynamics and average values, 47
Third Law introduced, 9
Thomson, William, Lord Kelvin, 4
tropical butterflies, flashing rate, 124
turbine, for diesel engine, 101
  introduced, 102
turbo-supercharging, 101
two-stroke engines, 100

U, internal energy, 203
units, 201

conversion table, 202
universal motor, Carnot's view of the
  steam engine, 1
universe and Universe, 30
  models introduced, 50
UP-ness, 143

w
  introduced, 65
  work, 205
water, role in dissolving, 158
water, structure, 159
Watt, James, 15
watt, W, unit, 203

work
  as area on indicator diagram, 19
  as method of transfer of energy, 23
  as stimulation with coherent motion,
    48
  as structure, 198
  definition, 14
  maximum available, 168
  produced by cell, 176
  production by chemical reaction, 167

Zeroth Law, 8
  introduction of temperature, 29

ABOUT THE AUTHOR

P. W. Atkins is a lecturer in physical chemistry at the University of Oxford and a fellow of Lincoln College. Both his B.Sc. and Ph.D. were awarded at the University of Leicester, followed by postdoctoral work at the University of California at Los Angeles. He was awarded a D.Sc. at the University of Utrecht in 1993. His research interests center on quantum theory, particularly the theory of molecular properties, which spawned his acclaimed book, *Molecular Quantum Mechanics*. He is also the author of *The Creation*, two other books in the Scientific American Library, and three widely read textbooks: *Physical Chemistry*, *Inorganic Chemistry*, and *General Chemistry*.

Selected hardcover books in the Scientific American Library Series

SUN AND EARTH
by Herbert Friedman

ISLANDS
by H. William Menard

DRUGS AND THE BRAIN
by Solomon H. Snyder

EYE, BRAIN, AND VISION
by David H. Hubel

SAND
by Raymond Siever

SLEEP
by J. Allan Hobson

FROM QUARKS TO THE COSMOS
by Leon M. Lederman and David N. Schramm

SEXUAL SELECTION
by James L. Gould and Carol Grant Gould

A JOURNEY INTO GRAVITY AND SPACETIME
by John Archibald Wheeler

SIGNALS
by John R. Pierce and A. Michael Noll

BEYOND THE THIRD DIMENSION
by Thomas F. Banchoff

THE SCIENCE OF WORDS
by George A. Miller

ATOMS, ELECTRONS, AND CHANGE
by P. W. Atkins

VIRUSES
by Arnold J. Levine

DIVERSITY AND THE TROPICAL RAINFOREST
by John Terborgh

STARS
by James B. Kaler

EXPLORING BIOMECHANICS
by R. McNeill Alexander

CHEMICAL COMMUNICATION
by William C. Agosta

GENES AND THE BIOLOGY OF CANCER
by Harold Varmus and Robert A. Weinberg

SUPERCOMPUTING AND THE
TRANSFORMATION OF SCIENCE
by William J. Kaufmann III and Larry L. Smarr

MOLECULES AND MENTAL ILLNESS
by Samuel H. Barondes

EXPLORING PLANETARY WORLDS
by David Morrison

EARTHQUAKES AND GEOLOGICAL DISCOVERY
by Bruce A. Bolt

THE ORIGIN OF MODERN HUMANS
by Roger Lewin

THE EVOLVING COAST
by Richard A. Davis, Jr.

THE LIFE PROCESSES OF PLANTS
by Arthur W. Galston

IMAGES OF MIND
by Michael I. Posner and Marcus E. Raichle

THE ANIMAL MIND
by James L. Gould and Carol Grant Gould

MATHEMATICS: THE SCIENCE OF PATTERNS
by Keith Devlin

Scientific American Library books now available in paperback

POWERS OF TEN
Philip and Phylis Morrison and the Office of Charles and
Ray Eames

THE DISCOVERY OF SUBATOMIC PARTICLES
by Steven Weinberg

THE SCIENCE OF MUSICAL SOUND
by John R. Pierce

MOLECULES
by P. W. Atkins

THE SECOND LAW
by P. W. Atkins

THE NEW ARCHAEOLOGY AND THE
ANCIENT MAYA
by Jeremy A. Sabloff